AF807432

# Progress in Lubrication and Nano- and Biotribology

# Progress in Lubrication and Nano- and Biotribology

Edited by
Catalin I. Pruncu, Amit Aherwar, and
Stanislav Gorb

CRC Press
Taylor & Francis Group
Boca Raton London New York

CRC Press is an imprint of the
Taylor & Francis Group, an **informa** business

MATLAB® is a trademark of The MathWorks, Inc. and is used with permission. The MathWorks does not warrant the accuracy of the text or exercises in this book. This book's use or discussion of MATLAB® software or related products does not constitute endorsement or sponsorship by The MathWorks of a particular pedagogical approach or particular use of the MATLAB® software.

First edition published 2022
by CRC Press
6000 Broken Sound Parkway NW, Suite 300, Boca Raton, FL 33487-2742

and by CRC Press
2 Park Square, Milton Park, Abingdon, Oxon, OX14 4RN

© 2022 Taylor & Francis Group, LLC

CRC Press is an imprint of Taylor & Francis Group, LLC

Reasonable efforts have been made to publish reliable data and information, but the author and publisher cannot assume responsibility for the validity of all materials or the consequences of their use. The authors and publishers have attempted to trace the copyright holders of all material reproduced in this publication and apologize to copyright holders if permission to publish in this form has not been obtained. If any copyright material has not been acknowledged please write and let us know so we may rectify in any future reprint.

Except as permitted under U.S. Copyright Law, no part of this book may be reprinted, reproduced, transmitted, or utilized in any form by any electronic, mechanical, or other means, now known or hereafter invented, including photocopying, microfilming, and recording, or in any information storage or retrieval system, without written permission from the publishers.

For permission to photocopy or use material electronically from this work, access www.copyright.com or contact the Copyright Clearance Center, Inc. (CCC), 222 Rosewood Drive, Danvers, MA 01923, 978-750-8400. For works that are not available on CCC, please contact mpkbookspermissions@tandf.co.uk

*Trademark notice*: Product or corporate names may be trademarks or registered trademarks and are used only for identification and explanation without intent to infringe.

ISBN: 978-0-367-49395-0 (hbk)
ISBN: 978-0-367-56110-9 (pbk)
ISBN: 978-1-003-09644-3 (ebk)

DOI: 10.1201/9781003096443

Typeset in Times
by codeMantra

# Contents

# Preface

Tribology is a multidisciplinary science that encompasses mechanical engineering, materials science, surface engineering (surface coating, surface modification, and surface topography analysis), lubricants, and additives chemistry. Since its early time, tribology – rubbing science – has been seen as an emergent and very challenging topic because it brings together different aspects as principles of friction, lubrication, and wear. The knowledge of tribology science was applied first about 3500 BC at a basic level for moving stone in which water was used as a lubricant in order to reduce friction and protect the surface of materials finally leading to reduced wear. Nowadays, the science of tribology is used very broadly and covers the fields of metals, polymers, ceramics, and a combination of composites with tremendous applications. To further help the research community and industry environment to be informed with the latest advancement, the proposed book will provide an update of all the different pairs of materials used in tribological contact in a widespread way. Leading researchers from industries, academia, government, and private research institutions across the globe will benefit from the knowledge provided by this highly application-oriented book. Moreover, it provides cutting-edge research from around the globe on the tribology field. Current status, trends, future directions, opportunities, and so on are discussed in detail, making it friendly for young researchers too. Furthermore, this part of the book is focused mainly on providing a systematic and comprehensive account of the recent progress in lubrication and nano-biotribology.

The contents of this book are spread over nine chapters.

Chapter 1 gives a brief overview of the recent advancement in biotribology followed by a careful discussion of its future projection. The major areas of biotribology research, such as joint, skin, and dental, are considered.

Chapter 2 focuses on the preparation methods and tribological performances of the best-known lubricant nanoadditives. The synthesis and modification processes of typical nanomaterials are described, and the rheology performance of lubricants with nanoadditives is introduced.

Chapter 3 discusses the recent progress in ceramic matrix high-temperature self-lubricating materials, including oxide ceramic matrix, nitride ceramic matrix, carbide ceramic matrix, boride ceramic matrix, and MAX phase ceramic matrix high-temperature self-lubricating materials.

Chapter 4 presents the current trends in the development of lubricating grease and its lubrication behavior.

Chapter 5 aims to inquire and foresee the operative analytical behavior of finite hydrodynamic bearing in the turbulence regime, which deals with non-Newtonian lubricants. The classical momentum and continuity equations have been utilized under turbulent and non-Newtonian flow, and details of the steady-state characteristics are presented too.

Chapter 6 focuses on the current development in the research area of bio-functionalized macro-porous Ti, which is a promising implant material. Some highlights are indicated for further efforts to study essential long-time clinic trials and applications.

Chapter 7 provides an overview of structural, experimental, and numerical studies on interactions between the ventral surface of snake skin and various substrates. Finally, some biomimetic implications of these results and future perspectives of studies on snake skin tribology are discussed.

Chapter 8 explores the tribo-behavior of Ti6Al4V (influential factors and interactive responses) and the ability of the EDM process to develop protective surface layers over the substrate used in the biomedical field.

MATLAB® is a registered trademark of The MathWorks, Inc. For product information, please contact:

The MathWorks, Inc.
3 Apple Hill Drive
Natick, MA 01760-2098 USA
Tel: 508-647-7000
Fax: 508-647-7001
E-mail: info@mathworks.com
Web: www.mathworks.com

# Editors

**Dr. Catalin I. Pruncu** is a Research Fellow in the Design, Manufacturing, and Engineering Management at the University of Strathclyde, Glasgow UK with 10 years of research experience in academia and industry. He has published more than 100 papers in ISI journals, 3 books, a patent, and other papers at various national and international conferences. Catalin is a Charter and Member of the Institute of Mechanical Engineers (UK) since November 2015. He has experience in prestigious universities (Imperial College London, University of Birmingham, University of Sussex) and industries such as IMI Truflo Marine Ltd. and Spanish Navy. Recently, he was invited as Editor for Special Issue, Wear Behavior of Polymer Composites and Mathematical Modeling and Simulation in Mechanics and Dynamic Systems, MDPI, and he is also a reviewer for almost 50 ISI journals including *Measurement, Elsevier, Journal of Materials Research and Technology, Surface and Coatings Technology, Journal of Cleaner Production*, and so on. He was involved in organizing different international conferences including the 12th International Conference on New Trends in Fatigue and Fracture, Brasov, Romania 2012.

**Dr. Amit Aherwar** is an Assistant Professor at the Department of Mechanical Engineering, Madhav Institute of Technology & Science, Gwalior, Madhya Pradesh, India. He received his Ph.D. from Malaviya National Institute of Technology, Jaipur, Rajasthan, India. He has more than eight years of teaching and research experience. His research interest includes tribology, biomaterials, surface characterization, multi-material and advanced composites, recycle/reuse of industrial wastes for engineering applications, and multi-criteria optimization. He has published more than 35 technical papers in reputed national and international journals/conferences, and also served as a reviewer for various journals.

**Prof. Dr. Stanislav Gorb** is a Professor and Director at the Zoological Institute of the Kiel University, Germany. He received his Ph.D. degree in zoology and entomology at the Schmalhausen Institute of Zoology of the Ukrainian Academy of Sciences in Kiev (Ukraine). Gorb was a postdoctoral researcher at the University of Vienna (Austria), a research assistant at the University of Jena, and a group leader at the Max Planck Institutes for Developmental Biology in Tübingen and for Metals Research in Stuttgart (Germany). Gorb's research focuses on morphology, structure, biomechanics, physiology, and evolution of surface-related functional systems in animals and plants, as well as the development of biologically inspired technological surfaces and systems. He received the Schlossmann Award in Biology and Materials Science in 1995, the International Forum Design Gold Award in 2011, and the Materialica "Best of" Award in 2011. In 1998, he was the winner of BioFuture Competition for his works on biological attachment devices as possible sources for biomimetics. Gorb is a corresponding member of the Academy of the Science and Literature Mainz, Germany (since 2010) and member of the National Academy of Sciences Leopoldina, Germany (since 2011). Gorb has authored several books, more than 500 papers in peer-reviewed journals, and 4 patents.

# Contributors

**Jun Cheng**
State Key Laboratory of Solid Lubrication
Lanzhou Institute of Chemical Physics, Chinese Academy of Sciences
Lanzhou, China
and
Center of Materials Science and Optoelectronics Engineering
University of Chinese Academy of Sciences
Beijing, China

**A. I. Costa**
CMEMS-UMinho—Center of MicroElectroMechanical Systems
Universidade do Minho
Guimarães, Portugal
and
Faculty of Engineering, DEMM—Department of Metallurgical and Materials Engineering
University of Porto
Porto, Portugal

**J. Géringer**
IMT Mines Saint-Etienne, Centre CIS, INSERM SainBioSE U1059
Université de Lyon
Saint-Etienne, France

**Elena V. Gorb**
Department of Functional Morphology and Biomechanics
Zoological Institute, Kiel University
Kiel, Germany

**Stanislav N. Gorb**
Department of Functional Morphology and Biomechanics
Zoological Institute, Kiel University
Kiel, Germany

**A. P. Harsha**
Department of Mechanical Engineering
Indian Institute of Technology (Banaras Hindu University)
Varanasi, India

**Basil Kuriachen**
National Institute of Technology Mizoram
Aizawl, Mizoram, India
and
National Institute of Technology Calicut
Kattangal, Kerala, India

**Jibin T. Philip**
National Institute of Technology Mizoram
Aizawl, Mizoram, India
and
Amal Jyothi College of Engineering
Kanjirappally, Kerala, India

**Shashank Poddar**
Department of Metallurgical and Materials Engineering
National Institute of Technology Raipur
Raipur, Chhattisgarh, India

**Sooraj Singh Rawat**
Department of Mechanical Engineering
Indian Institute of Technology (Banaras Hindu University)
Varanasi, India

**Sudip K. Sinha**
Department of Metallurgical and Materials Engineering
National Institute of Technology Raipur
Raipur, Chhattisgarh, India

**Sandeep Soni**
Tribology Laboratory, Department of Mechanical Engineering
S.V. National Institute of Technology
Surat, Gujarat, India

**Tianyi Sui**
School of Mechanical Engineering
Tianjin University
Tianjin, People's Republic of China

**Qichun Sun**
State Key Laboratory of Solid Lubrication
Lanzhou Institute of Chemical Physics, Chinese Academy of Sciences
Lanzhou, China

**F. Toptan**
CMEMS-UMinho – Center of MicroElectroMechanical Systems
Universidade do Minho
Guimarães, Portugal
and
IBTN/Br – Brazilian Branch of the Institute of Biomaterials, Tribocorrosion, and Nanomedicine
UNESP
Bauru, São Paulo, Brazil

**Dan Wang**
State Key Laboratory of Solid Lubrication
Lanzhou Institute of Chemical Physics, Chinese Academy of Sciences
Lanzhou, China
and
Center of Materials Science and Optoelectronics Engineering
University of Chinese Academy of Sciences
Beijing, China

**Jun Yang**
State Key Laboratory of Solid Lubrication
Lanzhou Institute of Chemical Physics, Chinese Academy of Sciences
Lanzhou, China
and
Center of Materials Science and Optoelectronics Engineering
University of Chinese Academy of Sciences
Beijing, China

**Shengyu Zhu**
State Key Laboratory of Solid Lubrication
Lanzhou Institute of Chemical Physics, Chinese Academy of Sciences
Lanzhou, China
and
Center of Materials Science and Optoelectronics Engineering
University of Chinese Academy of Sciences
Beijing, China

# 1 Advancements in Biotribology

*Sudip K. Sinha*
National Institute of Technology

## CONTENTS



## 1.1 INTRODUCTION

The science of 'tribology' primarily deals with the action of rubbing, friction, or wear and tear. The term tribology first coined from the Greek expressions τρίβω (tribó) meaning "I rub" and λογία (lógos) meaning "study or knowledge of" deals with the science of mutually contacting layers in relative motion in a tangential manner and includes the studies of friction, wear, and lubrication.

DOI: 10.1201/9781003096443-1

Historical evidence confirms that a base level of understanding on friction can be found even as long as 1493, when eminent personalities like Leonardo da Vinci first mentioned the underlying 'laws' of friction [1].

Da Vinci's postulates describe that the frictional resistance for any two dissimilar objects of identical mass will be the same, but the contact surface varies over distinct widths and lengths. Also, his theory states that the force required to overcome friction enhances twice when the applied weight doubles.

The field of tribology is very much interdisciplinary in nature and covers various subjects ranging from basic sciences, e.g. physics or chemistry, to mechanical or materials engineering. The economic importance of this multifaceted area of interest is immense since a wide range of products are damaged or destroyed because of tribological problems.

For example, the most usual example is in the application of bearing design. In addition to the manufacturing industry, other industrial products based on electronics or biomedical engineering also deal with tribological aspects.

Biotribology, on the other hand, is a compound word comprising 'biology' and 'tribology'. It is an emerging field of interest that investigates the phenomenon of wear and friction under biologically relevant conditions. It was first introduced by Dowson and Wright in 1970 [2] while working on tribology-related characteristics on biological systems. This fascinating, exciting, and extremely significant interdisciplinary research field primarily focuses on natural and artificial articular joints, prostheses and biological implants, ocular tribology, skin tribology, soft matter oral tribology, and others. On the other hand, areas concerning 'nanotribology' and 'green tribology' have come into recent existence in the field of tribological matters owing to their feasibility as environmentally friendly biomaterials for natural balance, stability on cost-effectiveness, and energy-conserving ability.

Figure 1.1a displays the vast online search outcomes on published articles, conference papers, books and book chapters, and so on dedicated to the field of biotribology

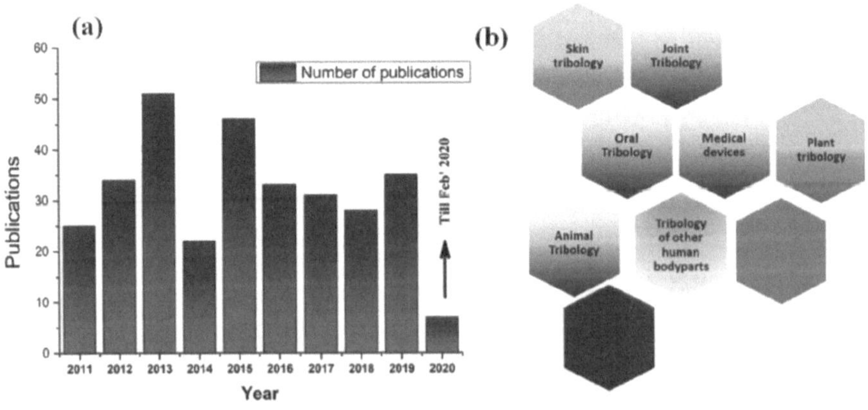


**FIGURE 1.1** (a) Extensive internet search results of relevant documents on biotribology, based on the Scopus database for the past one decade. (b) The schematic image shows contributions arising from the various domains of biotribology.

during the last 10 years. The immediate consequence of web search as found in the Scopus database on February 2020 yields a total count of '484' documents available in this area, which includes research articles, conference papers, review papers, book chapters, books, editorials, and so on (the keyword used for the search was "biotribology"). An average of 30+ publications can be found in this field within the last 10 years or so.

This chapter aims at the most recent advances in biotribological questions pertaining to the human body.

## 1.2 ARTICULAR TRIBOLOGY

In its early biotribological studies, joint tribology of synovial joints has been vastly inspired by mechanical bearing or similar mechanisms. Essentially, the field of articular or joint biotribology is controlled by the correct planning, fabrication, and execution of several forms of total joint (hip, knee, and so on) replacements. The significance of the field arises since the wear of contacting surfaces in living animals or humans can lead to unbearable pain and subsequently inhibit the motions concerning these body parts.

The usual concerns of extreme wear of the dissimilar materials in synovial joints cannot be neglected. Therefore, modern prosthesis development is totally based on the extensive use of biocompatible materials that are sufficiently robust and durable to survive in dynamic lifestyles of orthopedic patients. The primary reason arises as it minimizes the wear debris in such locations leading to inflammation and sufferings. Without the use of such biocompatible material arrangement, the component eventually collapses due to loosening of the prosthesis from its transplant position. In this connection, it must be mentioned here that the size distribution, primary shape, and cumulative volume of various wear particles are extremely crucial. For instance, while considering metal-on-metal hip implant bearings, the possibilities of formation of wear volume are usually significantly shrunk in comparison to polyethylene-on-metal implants. However, other factors such as harder metallic wear particles might prove to be a drawback for a metal-based prosthesis. Figure 1.2 shows the schematic and real X-ray images of a virgin and implanted hip joint in a patient.

### 1.2.1 Natural Synovial Joints

Articular joints in humans or other mammalian species suffer from sustained dynamic loading conditions (often cyclic in nature) during their entire life span. The bone structures in such joints are coated with a porous, smooth, and resilient tissue termed 'cartilage,' which resembles a rubbery pad-like substance. The cartilage is submerged and percolated with synovial fluid.

Natural synovial fluid is a non-Newtonian viscoelastic biofluid that saturates the void between two articulating bones to reduce friction. This matter apparently forms by a clear, sticky, gray, or straw-colored fluid which comprises a small number of (usually mononuclear) cells (~200 leukocytes per $mm^3$).

Synovial fluid or synovia comprises hyaluronic acid, which is the primary reason for its excessive viscosity.

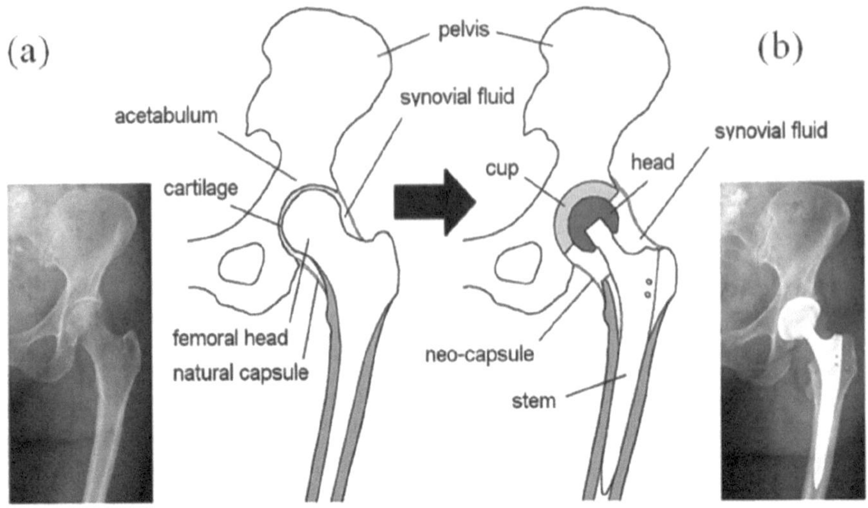


**FIGURE 1.2** (a) Naturally occurring hip articulation system offering articulation between the pelvic acetabulum and the head of the femur; (b) total hip replacement (THR). (Reproduced by permission of PLOS ONE publisher [3].)

The overall advantage of such naturally occurring biomechanical lubrication is to create interfaces with extremely low abrasion or friction [4].

In addition to the minimal friction, synovial joints show exceptional physical characteristics such as superior loading ability, shock tolerance capacity, strength, flexibility, and smooth movement with extended durability even in challenging situations involving severely loaded hip, knee, and ankle joints (also called talocrural articulation).

A schematic description of synovial fluid based on histological [5] and biophysical observation is depicted in Figure 1.3. It is to be mentioned that such structure has been produced under ex vivo conditions on many other occasions.

In a typical synovial joint, an articular cartilage acts as a lubricating object which covers the surface over the bones in the hip, knee joints, and so on to reduce the friction and wear between two hard surfaces while in motion. The coefficient of friction (COF) of a typical synovial joint is between 0.005 and 0.04 [6], which is significantly smaller than that of any standard engineering bearing surfaces.

### 1.2.2 Biotribology in Knee and Hip Joints

It is now known that from a clinical point of view, total joint arthroplasty (TJA) is one of the most prosperous and economical surgical techniques being accomplished as a potential cure for last-stage osteoarthritis. However, gradual discharge of polyethylene wear particles and metal ions from implant structures may deteriorate their final properties, which is further enhanced by biological reactions in the nearby regions.

The concept of TJA was first conceived by Sir John Charnley who first introduced [7] low friction cemented arthroplasty for total hip joint replacement during the late

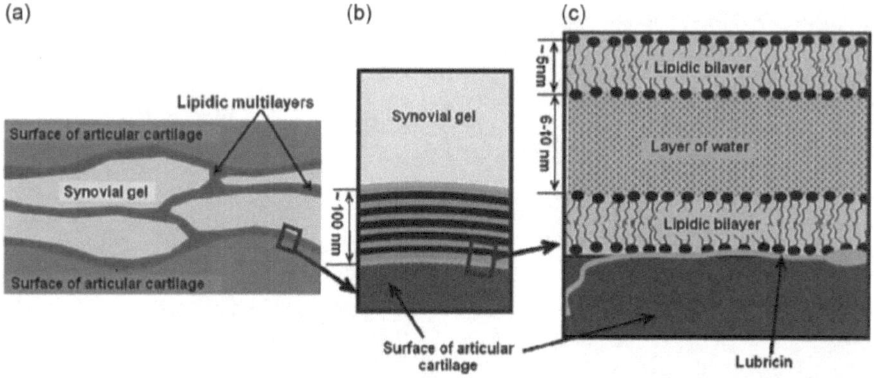


**FIGURE 1.3** Graphic representation of synovial fluid: (a) Lipidic multilayer sacks made up of synovial gel in an irregular manner; (b) multilayered lipidic assembly at the common boundary among the synovial gel and tendon; and (c) microscopic view of the lipidic multilayered assembly attached on articular cartilage by lubricin. (Reprinted with permission from Taylor & Francis [5].)

1960s, which is, to date, considered the gold standard in this section. His unique design primarily consists of three components: a stainless steel-based femoral head, an acetabular cup that was built up of ultra-high-molecular-weight polyethylene (UHMWPE), and the essential bone cement or the synovial fluid that was aimed at lubricating this ball-on-socket type joint in order to reduce its friction.

His design attributed to minimizing the joint friction, in addition to the torque developed by this rubbing action between the dissimilar components. The initial materials used by Charnley for the joint combination were based on PTFE and stainless steel as PTFE offers the least possible frictional coefficient among solid bodies.

In comparison to Sir Charnley's discovery, G.K. McKee and J. Watson-Farrar designed and fabricated a metal-on-metal (MoM)-based joint prosthesis for the application of total hip replacement (THR) [8]. These MoM hip implants have been incorporated into patients with initial success. However, within a time span of 10 years or so, these artificial implants show clear and distinctive indication of aseptic loosening, which eventually leads to implant failure. It is believed that the failure of such implants is not solely due to the faulty material combination, but also arises owing to the inferior implant design.

Then onwards (1972), the low-friction Charnley metal on polyethylene prosthesis started to dominate the orthopedic market.

Metallosis is hypothesized to a medical condition involving infiltration of articular soft tissues and/or bones by metallic deposition arising due to wear debris due to a failed prosthetic implant in areas like knee joint, hip joint, shoulder, elbow wrist, and so on.

The frequency of such occurrence is around ~5% of all prosthetic implants in the last 40 years. In general, women are more vulnerable to such risks compared with men.

Verde et al. [9] investigated the medical condition of a 45-year-aged man who was suffering from osteosarcoma of the distal femur (knee joint) in association

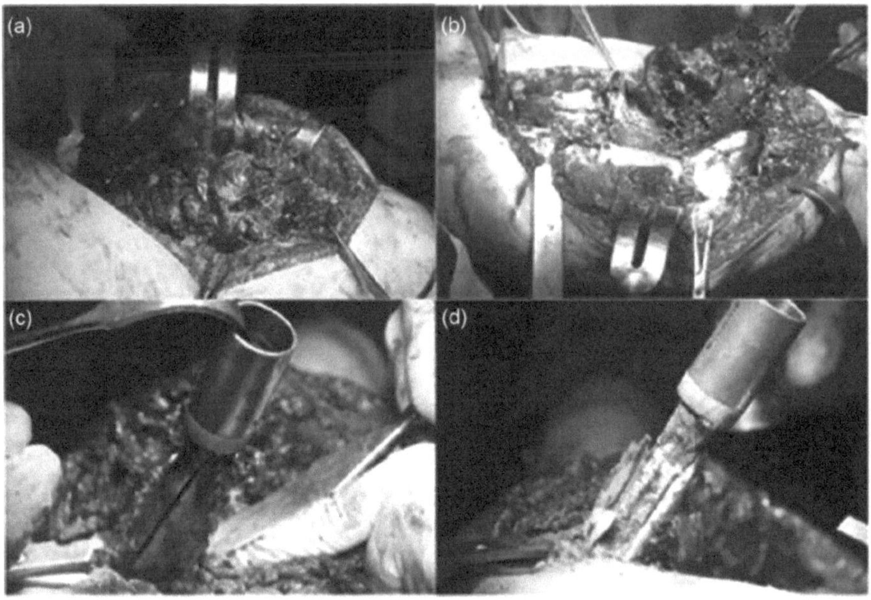


**FIGURE 1.4** (a–b) Surgical photographs display enormous soft tissue metallosis of the adjoining to the prostheses; (c–d) the femoral prosthetic stem strongly unified with the bone structure, lacks any osteolytic reaction. (Reprinted with permission from Taylor & Francis [9].)

with a modular prosthesis when the patient was only 18. Seventeen years later, the patient started suffering from massive metallosis with a clear symptom of skin dyspigmentation.

The authors established the existence of enormous soft tissue metallosis of the distal femur as well as proximal tibia, as confirmed by a revision knee surgery (Figure 1.4a and b). Figure 1.5 shows both femoral and medial tibial plateau fractures once the prosthesis components were removed. It is found that a large amount of polyethylene wear and distortion were distributed unevenly onto the medial and lateral joint surfaces, and these are the primary reasons for such wear debris formation.

This investigation revealed that if prosthesis fracture is identified, surgical procedures should be tried as soon as feasible.

It is now well understood that for developments in the design aspect of knee arthroplasty, selection of novel materials, sterilization scheme corrosion resistance, and new techniques in in-vitro wear simulation have led to property enhancement of total knee prostheses by eliminating the wear rate, fatigue, delamination, and other mechanochemical effects [10–12]. Considering the above-mentioned points, it is imperative to understand that decreasing wear rate in joint regions, in addition to the development of new-generation bearing materials and superior implant design, is a necessity while applying knee arthroplasty on various categories of patients ranging from younger to aged ones or heaver to more dynamic patients in the present era.

Adverse surface sensitivity for reactions of metal ion species in local tissue reactions was observed in patients suffering from complications with a higher value of

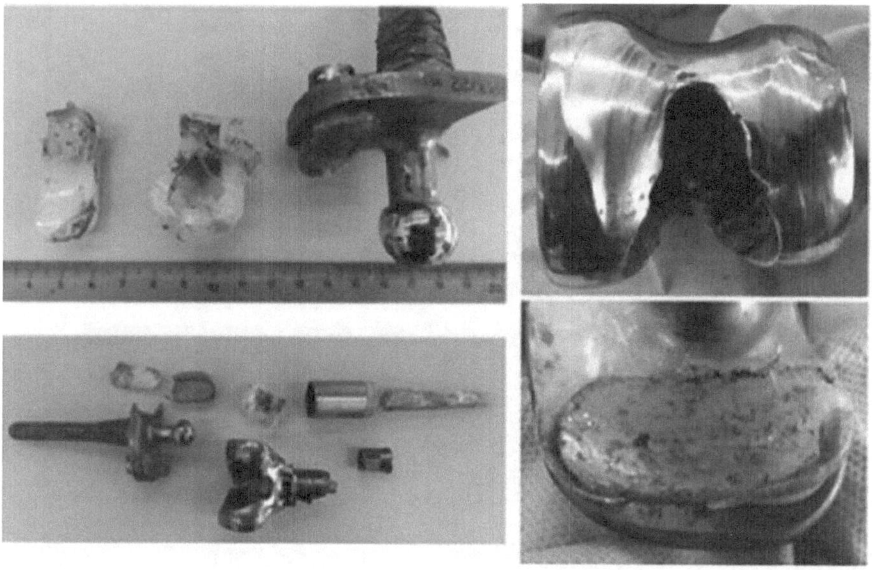

**FIGURE 1.5** Severe damage of prosthesis components of the femoral stem and medial tibial plateau owing to massive polyethylene wear and distortion around the medial and lateral joint faces. (Reprinted with permission from Taylor & Francis [9].)

cobalt retention in blood. The problem of Co-ion leaching from the CoCrMo substrate material in patients undergone total knee implants can be alleviated using a multilayer coating of zirconium nitride (ZrN) as demonstrated by Reyna et al. [13]. Such ensembles have the potential to prevent Co-ion release metallic substrate and thereby decrease the material removal rate of the UHMWPE liner when matched with a bare implant. Typically, the thickness of the ZrN multilayer coating deposited on the CoCrMo substrate ranges between 3.5 and 5.0 μm. It is found from their analysis that the wear rate can be substantially reduced by the ZrN group ($1.01 \pm 0.29$ mg/million cycles) as compared to the untreated CoCr group ($2.89 \pm 1.04$ mg/million cycles). Also, the surface layer of ZrN-coated femurs exhibits a shining appearance and smooth texture once the trial is completed. Wear scratch marks can be observed onto the untreated femur surface (Figure 1.6). Another notable outcome of this research effort is the drastic reduction in metal ion discharge from the ZrN-coated knee joint transplants, which ranges up to three orders of magnitude compared to the bare CoCr implants.

## 1.3 SOFT MATTER AND SKIN TRIBOLOGY

Knowledge of friction and lubrication of human skin is critical since it plays a significant role in the tribology of skin and related soft body parts in everyday life. For instance, while grasping a sports component or experiencing a lotion or ointment for medical or cosmetic purposes, the body part undergoes a rubbing action leading to a palpable effect on it.

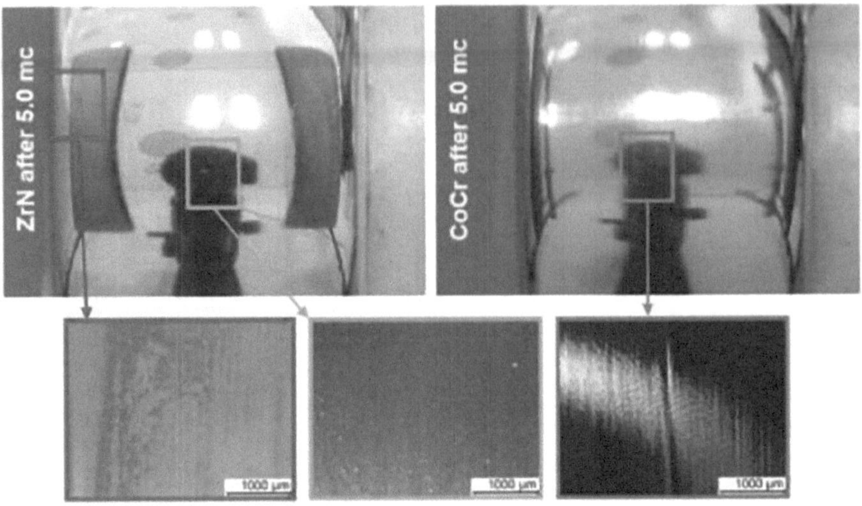


**FIGURE 1.6** Macroscopic images showing wear characteristics of the femur components (with and without ZrN layer deposition) after 5 million HDA cycles, confirmed by the hazy image of the camera and lights [13].

Microscopic investigations of an individual's skin have revealed that it has a complex structure that comprises a system offering uniform tensions by a flexible collagen tissue in the outermost dermis. This outermost protective layer is created at the onset of birth, and it increases in size and depth, with age until adolescence time. The nerves and associated sensory components act as pain detection, whereas the surface, temperature, and individual glands are contributory in the formation of sweat and sebum. In general, the skin is a mixed, anisotropic, cemented material displaying an irregular stress–strain curve. Its complex structure provides mechanical features which are often difficult to interpret [14,15]. This information is important for cosmetic and medical research.

A schematic description of human skin is shown in Figure 1.7. As already mentioned, the surface layer of the human body is called the epidermis. This layer is labeled as the stratum corneum, which is primarily composed of some dead cell tissue.

This layer is comparatively much stronger and harder with a thickness of ~10 μm [16,17]. However, it is much thicker in some areas like the bottom of the feet. Next to the dermis, another relatively thicker layer, known as the epidermis, is present. This layer is comprised of connected fibers in a semi-fluidic matrix. Beneath the epidermis, another fat-based cutaneous tissue is located which has a high fluid content of ~60%–70% by volume [18].

### 1.3.1 Friction of Human Skin

As already stated, skin tribology is directly associated with the use of cosmetic and skincare products, or consequences of dermatological issues involving skin disorders or injuries, maturity, wound restoration, and artificial implantations.

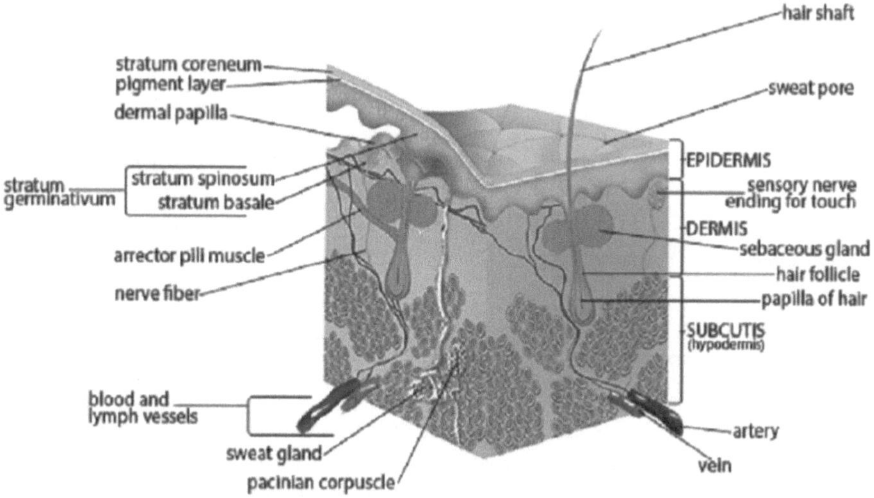


**FIGURE 1.7** Schematic representation of the various functional layers of human skin [19].

The amount of moisture content or water plasticization in the stratum corneum (SC) of the skin greatly affects the frictional properties of our skin [20]. It is also aggravated by the water content in the periphery between the skin and the subcutaneous layer [21].

Researchers have ensured that human skin, on most occasions, shows a COF of 0.1–1.3 [22]. The COF value typically drops as the applied load enhances, particularly in small load regimes [23,24]. It is believed that there are two mechanisms responsible for the inherent friction of human skin. The first one is 'plowing' which generates frictional force owing to the mechanistic coupling of substances that naturally arises from surface roughness factors.

The second reason arises from the required load that is used to overwhelm adhesion between two mutually inclusive surfaces. In this case, atoms are removed from the second material layer onto the initial material's surface, and this process subsequently generates frictional forces.

Zahouani et al. [25] have carefully exercised a series of in-vivo studies involving macro-indentation and tribological assessments on 46 cases in the age group of 18–70 years separated into three units. The elasticity of the skin diminishes as the loss factor rises with maturity, as evident from indentation tests. A complex modulus, ranging from $7.17 \pm 2.06$ kPa for the eldest class to $10.7 \pm 2.64$ kPa for the teenage class of people, is obtained after the dynamic indentation tests, evaluated at 10 Hz frequency stress.

### 1.3.2 Hydration of Skin

The COF of human skin depends on several external and internal factors once they make a contact pair with the external body.

**TABLE 1.1**
**Influence of Hydration and Temperature on Human Skin [27]**

| | Skin Temperature (in °C) | | | Skin Hydration (in AU) | | |
|---|---|---|---|---|---|---|
| | Range | Median | (IQR) | Range | Median | (IQR) |
| Total | 21.4–35.1 | 32.1 | (2.7) | 4–112 | 25.5 | (3.1) |
| Ventral forearm | 29.1–34.9 | 32.2 | (2.6) | 13–56 | 41.0 | (10.0) |
| Dorsal forearm | 30.7–34.6 | 32.4 | (1.7) | 4–47 | 29.0 | (19.3) |
| Index finger pad | 21.4–35.1 | 30.2 | (6.2) | 22–112 | 78.4 | (47.0) |
| Dorsum hand | 30.3–33.2 | 32.1 | (2.8) | 10–43 | 25.5 | (3.1) |

'Hydration' of the skin is indisputably one of the most crucial of those external factors, since both physiological and ecological conditions have significance on it. From a practical standpoint, it is frequently realized in our daily life that there is an increase in friction between the contact pair [20,26], when the moisture or water content of the skin increases; e.g. in sports events, the gripping is much better when a fabric is in firm contact with sweated skin. The COF of skin usually varies between 1.5 and 7 under dry and wet environments. The hydration factor for various age and gender does not have pronounced effect since they do not have direct contribution to impact the COF. Veijen [27] in his thesis work has described the effect of skin temperature and hydration comprehensively. Table 1.1 describes his analysis in a nutshell.

His work defines that the variation of both the skin temperature and its moisture content is considerably distinct throughout the anatomical locations. Also, it can be clearly observed that the range of either skin hydration or skin temperature of the locale index finger pad is comparatively large when estimated in comparison to the other bodily locations.

The variation in friction in various anatomical regions, for instance, the finger pad and the palm, is significantly different due to the greater skin hydration. The reason behind hydration is that water, creams, and lotions/moisturizers have a significant role in controlling the elastic properties of the SC in addition to its adhesive nature and frictional attributes [28]. While in a hydrated situation, the COF enhances as resulting from the softening tendency of the SC and due to the rise of the adhesion stresses at the skin–body contact.

### 1.3.3 Contacting Materials and Their Effect on Skin

It is imperative to understand that the COF is not constant in a typical skin–material contact and differs considerably upon the working environments, atmospheric conditions, and selection of proper resources. It also possibly depends upon the nature of the movement that is subjected to an investigation. A group of researchers [29] have provided a comprehensive list of such effects on COF of human skin under variable conditions. Different studies have observed that the COF of fabrics against human skin is mostly affected by the type of the cloth, dampness of skin, and atmospheric moisture content. It is now well understood that various forms of skin injuries may readily occur once the human skin and textiles are in mutual contact for a prolonged period. These studies include, but are not limited to, the possible mechanism on how skin friction (various functional regions) differs, shearing action against several types of resource materials (i.e., clothes for medical usage) in various contact environments.

A study conducted by Derler et al. [30] shows that new hospital bed sheets are less vulnerable to human skin owing to their reduced interaction area, greater open interface area, and 1.5 times smaller shear strength. In a separate study, Gerhardt et al. [31] have investigated the effect of epidermal hydration on the friction and shear forces among a group of men and women's skin against textiles (Figure 1.8). Initially, the skin hydration and viscoelasticity of a group of selected persons were estimated by using standard corneometry and a suction procedure, respectively. It was found from their analysis that the friction of feminine skin revealed considerably greater moisture sensitivity. The COF was normally enhanced by 43% (females) and 26% (males) once the skin hydration fluctuated between extremely dry and naturally moist skin.

### 1.3.4 Effect of Sliding Velocity and Rotation

As is already said, skin plays an essential role in protecting the human body as an outermost layer. Therefore, understanding the various causes of the generation of

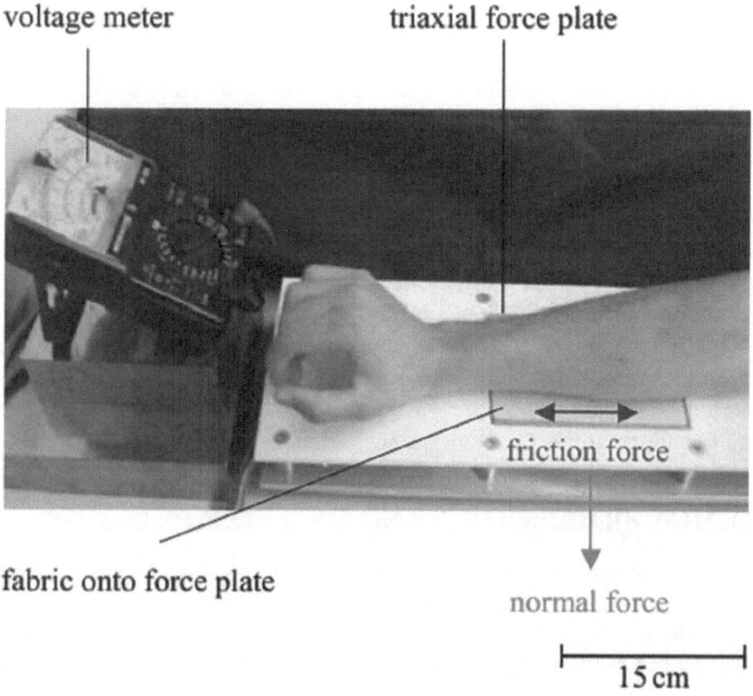


**FIGURE 1.8** Typical skin–fabric friction experimentations on a force plate under in-vivo conditions. (Reprinted with permission from Taylor & Francis [31].)

skin friction and its subsequent remedies plays an important role not only in clinical or skincare applications but also in garments or textile utilization and many other purposes associated with skin.

Researchers have found that there is a sizable plastic deformation when the asperities of the hard contact surface mutually interact with the softer skin surface. The rigid probe material slides on a viscoelastic skin surface and the deformation amount and the mechanism entirely depend on the skin anatomy.

The probe surface is always tougher than the skin layer underneath. Therefore, upon pressing the probe into the skin surface, these asperities located on the probe surface tend to infiltrate and directly interact with the skin layer, thereby producing physical impression or grooves in the probe–skin contact surface, if the shear stress is above the threshold limit.

Tang et al. [32] have demonstrated that sliding speed plays a significant role in the frictional performances of an individual's skin. According to their study, upon increasing the sliding speed from 0.5 to 4 mm/s, the coefficient of dry sliding friction is enhanced. In addition, the "stick-slip" tendency becomes more severe.

The sliding speed is directly responsible for the hysteretic friction increase, which is further accompanied by a major loss in the energy absorbed due to elastic hysteresis.

Zhang et al. [33] have shown in their study that the COF increases $7\% \pm 2\%$ once the speed of cyclic motion rises from 25 to 62.5 rpm.

Meredith et al. [34] have inspected the effect of various kinds of fabric materials and their dynamic COFs on skin experienced in abrasion injuries faced by motorcycle riders.

For this, sample riders were tested while they have been participating in artificial crashes (number = 92 riders).

The coefficient of dynamic friction of the artificial materials used by the motorcycle riders against an artificial skin alternative, named 'Lorica® Soft', was verified by applying a customized friction testing system involving a biaxial mechanism.

It is to be mentioned that 'Lorica® Soft' is a synthetic leather that comprises a polyamide fabric over which a polyurethane (PUR) coating is applied.

Their study could not establish any obvious relationship between the frictional coefficient (COF) between these two heterogeneous interfaces and any possibility of abrasion wounds to such accidental conditions. Their investigation indicates that while certain materials may cause greater amounts of abrasion damages, still the 'COF' might not be utilized to envisage such occurrences.

## 1.4 DENTAL BIOTRIBOLOGY AND ORAL PROCESSING

As already discussed, biotribology defines itself as the tribological phenomenon related to the bio-systems. Interestingly, it includes mammalians; however, it could be either the human body or animal body. Nevertheless, the story does not stop here. When it comes to the bio, the rubbing processes between two biological active surfaces may occur after implanting an artificial material in the human body and tribological processes naturally occurring in or on the tissues and organs of animals [35]. The study of the wear of teeth due to interaction in response to mechanical degradation and biological factors including salivary compositions in the oral environment due to oral mucosa is termed dental biotribology. Mammalian teeth are associated with the masticatory organ that is responsible for the movements inside the mouth. The occurrence of events like crushing, grinding, and squeezing involves continuous movement and falls in the category of dental biotribology [36]. Human teeth are classified into four different types, i.e. incisors, canine, premolars, and molars. All classified types are surrounded by human biological fluid saliva, which may present acidic components due to fermented carbohydrates originating from the decay of food that might trigger off dental plaque formation. In addition to this, while eating, saliva is known to be the most essential biological active fluid, which governs the oral environment and decides the fate of enamel. During intake, saliva encounters any liquid food or solid food and gives rise to the formation of a bolus. However, their study and interaction between those two are still ambiguous [37]. The rate of wear of dental materials due to attrition, adhesion, abrasion, and friction relies on saliva's biological and physiological functions and its oral mucosa. However, various factors such as eating habits, aging, dieting, and brushing affect the composition of saliva and it might interfere with the interaction between saliva and food components [38]. Hence, it is very important to understand the effect of oral hygiene and its activity against wear action to understand the dental implant materials in view of its oral behavior.

### 1.4.1 Wear of Human Teeth: Main Factors

Wear can be described as the gradual removal of material caused by mass and volume loss from the body due to rubbing action. Human teeth are a composite mixture that comprises dentin, enamel, dentin enamel junction, and pulp and each one of them has a significant impact. Enamel consists of both organic and mineral phases, and it is known to be the hardest tissue in the dental regime due to its anisotropic orientation of enamel rods. However, it is exposed at the occlusal surface. Dentin is pale yellow in color encircled with enamel, cementum, and pulp. It is highly mineralized with 70% HA with collagen type 1 and fluid rest. Moreover, it is a composite mixture consisting of dentin tubules, peritubular dentin, and a collagen matrix of intertubular dentin and hence is isotropic in nature. Dental enamel junction is an interface region between the dentin and outer enamel coating and is solely responsible for the prevention of cracks near the enamel due to its structural morphology and hidden biomechanical properties. Dentin pulp is a soft vascular tissue that helps teeth in their formation, provides nutrition, prevents cavities, and protects teeth from inflammation [39]. To understand the underlying wear mechanism that is intricate, however, certain variables affect the wear properties of teeth. Mutual interactions between the individual subjects like abrasion, oral pH, biological systems, chemical composition, food ingestion, tongue mucosa, drinking beverages, different foods, and mastication are all the several individual factors that are not solely responsible alone. However, understanding the tribology of the hierarchy of dental enamel is ambiguous and complex to understand the surface texture. The critical factors are classified into two groups: internal factors and external factors. Internal factors that are responsible for the surface loss are attrition, erosion, abrasion, fatigue, resorption, microbial factors, saliva, and its pH. External factors include fluoride concentration of toothpaste, beverages, and acidic foods. Most of the wear contents are due to mastification, which involves movement of two upper- and lower-jaw fluctuations associated with the voluntary muscles. All this happens in the presence of salivary mucus, which also acts as a lubricating media. Moreover, it helps to swallow the food in the presence of oral enzymes, which contain salivary amylase and lingual lipase coming out of salivary glands for pre-digestion and helps in swallowing [40]. Open-phase mastication does not involve the chances of a higher wear rate. Though, this is the first stage, when the upper and lower molar teeth are brought to a position of contacting the food bolus with the aid of occlusal forces. However, in a closed phase, basically the second stage, the chances of wear rate are higher as it involves the grinding and chewing of the food bolus. It happens when two molars of different jaws have to encounter immediate action of occlusal loads. These loads are distributed toward food bolus or particles trapped inside into the two molars. Thus, it might wear out the two opposing surfaces of teeth due to shredding and scratching during the mastication. In addition to this, there are two other phenomena that include thegosis and bruxism. These phenomena happen when the crown teeth of the upper and lower jaws are not aligned while past sliding, chewing, and eating food. This part gives more frictional forces to tear the thing that highly affects the surface of the teeth [41,42]. In thegosis, the sides of two upper- and lower-jaw molar teeth slide edgeways due to the involvement of direct contact surfaces. On the other hand, bruxism involves the grinding of teeth as a response to stress. It might happen due to anxiety or some psychological tensions and disorders resulting in unconscious clenching of teeth [43,44]. One of the governing factors includes the pH of the oral environment or simply we could say the saliva and its compositions that contain proteins, sugar, amylase, and so on. It has been reported that at lower pH values, the chances of wear increase. The acidic aqueous phase due to disturbed bowel movement and bad eating habits influenced the mouth pH that leads to the erosion of the enamel layer due to chemical inhomogeneity and loss of elastic strengths and other mechanical properties. Dental erosion entails chemical reactions between the enamel and dentin in the presence of a disturbing under saturated aqueous phase. Erosion is unlikely to weaken the surface texture of enamel and dentine. Moreover, taking the opportunity of mechanical factors like abrasion, attrition, and fatigue is accountable for surface wear of the teeth [45]. Other factors include microbial factors like bacteria, microorganisms, and so on. These tiny microorganisms are always present over the surface of teeth. If not given proper treatment and maintenance to teeth, pathological issues will take advantage. Due to this irregular poor habit, microtribology of surface starts to take place. Over a period, there will be partially wearing of teeth around the gum side that will start happening. Due to continuous grinding and crushing, the partial erosion eventually converted into final erosion and becomes the major cause of wear due to those microbial factors. Researchers have studied the dietary effect on surface wear in Mesolithic and Neolithic levels at people of Tell Abu Hureyra, Syria. The Mesolithic group was given a soft diet with small-grained seeds that needed a little occlusal load in grinding. While in the Neolithic group, large-grained cereals were given. It was found that a change in the diet between two groups could result in a microwear of teeth. In their study, it was concluded that food diet differences in two dissimilar groups that change in the grain diet have had the consequences to a greater extent on the microwear of surfaces than changes in the meat diet [46]. Patients with chronic alcoholism are affected by dental wear problems. It has been found out that those patients with higher consumption of alcohol have higher tooth wear due to attrition. In addition to this, it has also been suggested that this wear effect is next to bruxism in which alcohol stimulation of the brainstem reticuloactivatory system generates masseteric muscle contractions and bruxism with the brisk movement of eyes during sleep. Furthermore, it entails erosion due to no occluding surfaces resulting in different wear patterns [47,48].

Oral environment circumstances provide an opportunity to wear out the human teeth. There are several hidden factors too by which teeth are severely affected directly or might be indirectly. Gum problems, infection, microorganisms, types of food intake, aging, and accidents impair dental tribology. Dental caries or cavities, plaques formation due to a thin layer film (pellicle) over the teeth that contains macromolecules and lot of bacteria, and lack of maintaining oral hygiene are micro-level disguised factors that set off dental biotribology.

### 1.4.2 Effects of Metals and Alloys

For long, d-transition metals and their alloys have been widely and commonly used for dental prosthesis materials as well as restorative materials owing to their biocompatible properties. However, corrosion of these alloys in an oral environment is a critical issue. Elements like Au, Ti, Ag, Cd, Pd, Ti, Al, and related alloys are being used in dental material applications. However, iatrogenic exposures of these elements present in the alloys lead to complex anticipation with dental tissues. Hence, every material shows a different response to dental tissues. Thus, understanding the intricate behavior of those materials with the living tissue is being a great challenge to the researchers. Among various materials explored, commercially pure titanium (Cp-Ti) has gained attention in dental material research. Ti and its alloys are uniquely versatile materials used in medical implant application owing to their excellent biocompatibility and compatible elastic modulus in tissue response [49,50]. The lack of wear properties of Cp-Ti makes the material unsuitable for tribological applications. This can be overcome by adding alloying elements to make it more wear resistive as well as more compatible with body fluids. Iizima and co-workers studied the wear properties in between Cp-Ti and Ti-6Al-7Nb for dental casting as a prosthodontic application in a simulated occlusion wear load patterning test. Moreover, lower volume loss in the alloy has been identified in their study when compared to Cp-Ti. In addition to this, electron microscopy images revealed that the surface texture of the alloy was much smoother as compared to plane Ti. From their study, it has been concluded that the Ti6-Al-7Nb alloy is more appropriate to reduce the risk of failure due to occlusional load behavior [51]. Ti6Al4V, Ti6Al7Nb, Ti-5Al-2.5Sn, Ti6Al6V2Sn, and so on belong to the category of alpha-beta alloys present both in the alpha and beta phases. Aluminum is the principal alpha stabilizer that helps to strengthen the alpha phase. Beta stabilizers, such as vanadium, are also responsible for strengthening and permit these alloys to be hardened by solution heat treating and aging (STA). Alpha-beta alloys are a good combination of mechanical properties and strength. C. Ohkubo and co-workers had worked on the wear behavior of Cp-Ti and Ti6Al4V alloys by varying the copper concentration. In order to evaluate the wear resistance properties, Cu has been added into the system. It is to be believed that the introduction of copper leads to the formation of an alpha Ti/$Ti_2Cu$ eutectoid phase in alpha titanium. In addition, it modifies the microstructure by reducing the stacking fault energy that might be beneficial to enhance the wear properties in dental materials [52]. Apart from Ti and its alloys, there are some other dental alloys that have been used like wironium, cerapall-2, aurofluid-3, wiron-99, and pagalinor-2. Moreover, it has been found that the wear behavior of alloys with abrasive types based on precious metals such as Pd, Au, Ag, Pt and CoCr is higher in comparison to Ti alloys. [53]. Hisashi Do et al. worked out over a titanium-based Ti-5Al-13Ta alloy cast for dental implant applications. In their study, their research team has assessed the fatigue behavior of the developed alloy produced with the help of a centrifugal casting machine in comparison to Cp-Ti, Ti6Al4V, and Ti6Al7Nb in the presence of NaCl solution as media. Fatigue is also a part of wear and occurs due to fluctuation of stresses that are much lower than the stress required to cause failure. Moreover, the ratio between the fatigue limit and UTS was found to be 30% for Ti-5Al-13Ta when compared to Ti6Al4V and Ti6Al7Nb alloys. In addition, a low fatigue life with a fatigue limit of 220 MPa was observed when it was conducted in 0.9% sodium chloride solution. Hence, the authors suggested the lower risk of fatigue failure that indicates the optimistic role in the application of dental prosthesis [54]. Nitinol is a Ni–Ti alloy known for its shape memory effect and is the most common material used in orthodontics for arch wires applications. It has been evidenced out that higher amounts of nickel release into the body are harmful that might be allergic. In Ni–Ti alloys, the concentration of nickel is more than 40%; hence, the material can release excess Ni ions in application to the intraoral environment and might be hazardous. It has also been suggested that a concentration of Ni of around 30 ppm might influence the cytotoxic response. The leach-out of Ni ions in dental applications could happen due to mechanical action or might be due to oral mucosal injury, and all these activities take participation within the oral environment [53,55].

### 1.4.3 Effect of Ceramics

Ceramics are nonmetallic inorganic substances that are usually brittle in nature. It has been used from ancient times in the form of porcelain, pottery, glassware, and many more. Moreover, it has a quality to withstand at higher temperatures and is amorphous as well as crystalline. In the stress–strain curve, it only shows the failure up to an elastic limit. Nevertheless, it exhibits good wear properties along with fine texture color mimicking properties and excellent chemical resistivity; such attributes of ceramics have been proven fruitful to be used in medical implant applications as dental prosthetics [56]. However, researchers are trying to figure out the abrasiveness of the dental ceramic implant to understand its rubbing action against its counter-natural teeth. Zirconia is one of the oxide ceramics that are being used in dentistry for a period. It is basically oxide zirconium derived from the mineral zircon ($ZrSiO_4$). When compared to other biomaterial ceramics, this material has gained popularity due to its superior mechanical properties. Although it is bioinert in nature, its biocompatibility, higher elastic modulus, antibacterial properties, and lower tissue inflammation make this material a suitable candidate of choice in prosthetic dentistry [57]. Ioannidis et al. designed 3D printed ultra-thin zirconia veneers with the assistance of CAD/CAM software. In addition, they have also built milled zirconia and heat-pressed lithium disilicate using the CAD/CAM technology. The purpose of preparing these materials for load-bearing capacity is to gain a better insight into the mechanical properties of the artificial ceramics with a view to understand the applicability in wear properties of the dental materials. The median fractures at ($F_{ini}$) and ($F_{max}$) values obtained for 3D-printed CAD/CAM zirconia, CAD/CAM-milled zirconia, and heat-pressed lithium disilicate were found to be 1650N, 1250N, and 500N and 2026N, 1500N, and 1555N, respectively. From their results, 3D-printed zirconia has shown optimistic value against mechanical properties. Despite the fact with their differences in mechanical properties, these candidates could be potentially useful for dental prosthetic restoration [58]. Dental porcelain has been used for the past many years due to its esthetic characteristics together with good wear properties. Feldspathic porcelains are a kind of ceramics that has been widely used in the field of medical dentistry, although their brittle nature makes them a poor candidate for vast use in prosthetics application. Failure due to repeated occlusal activities and the unhealthy oral environment is a major concern when tested in vivo. Apart from this, the friendly nature of the body and its good texture properties make this material useful especially in veneer applications. H.Y. Yu and co-workers studied the friction/wear properties in two different varieties of feldspathic porcelain as opposed to $Si_3N_4$ balls in different loading parameters at a different frequency in the presence of artificial saliva as a lubricating agent. VMK 95 and CerecVitablocs Mark II are two different dental feldspathic porcelains that were chosen in their study. Thus, after experimentation, the wear behavior owing to the abrasiveness of Vita VMK95 was found to be more with a low frictional coefficient in comparison to CerecVitablocs Mark II. However, in the absence of a lubricating agent, CerecVitablocs Mark II is more prone to delamination and surface cracks due to brittleness, especially at lower sliding cycles, whereas lateral cracks are more favorable in higher sliding cycles. At 10N loads, the value of the COF in the presence of lubricating media was about 0.84, which was lower than the value in the dry state (about 1.2–1.5). Besides this, on increasing the load to 40N from 10N, the value of the steady-state coefficient declined to 0.84–0.54 [59]. Hence, it is well known that ceramics are brittle in nature. However, there is a limitation for the long-term use of bioceramics as dental prosthetics in the human body. Few ceramics got obsolete due to their weak mechanical properties and cannot be used in a run for a long time. Therefore, in the future, there will be a need for advanced bioceramics of improved mechanical properties with fine texture and good esthetics and high antibacterial quality along with the high compatibility to the human body. Not only this, but the material should also be economic and could be used for a long-term dental implant.

### 1.4.4 Effect of Composites

Composites are lightweight anisotropic materials made up of two or more dissimilar materials and are chemically homogeneous by nature. Their higher strength-to-weight ratios make them lighter with superior mechanical properties as compared to alloys and ceramics. Moreover, they have been used in the field of dentistry as dental restoration materials in the form of different fillers or amalgams as ceramic-based composites. However, due to several mechanical factors, there might be a chance of coming out of some particles from those dental composites. Resin- or polymer-based composites have taken attention due to their higher toughness, higher antibacterial properties, improved esthetic properties, and anti-corrosive properties [60]. Xin hu and co-workers studied the wear properties of different dental materials by investigating their wear attributes as opposed to counter steatite ceramic grinding wheel. The materials used in their study were P-60, Z350 resin-based composite and Au-Pd alloy-based dental materials. The experiment was conducted using a pin-on-disk apparatus in the presence of artificial media. Maximum wear loss has been identified in an Au-Pd alloy-based dental material as compared to a P-60 light-cured composite with low volume loss. Moreover, from the volume loss study, it was found that the wear loss was of an abrasive kind [61]. J.C.M. Souza et al. [62] inspected wear operation over four resin-based composites for dental restoration. Furthermore, artificial saliva and toothpaste were used as media in their study. A stainless steel ball with a 3 N load was used to study the abrasiveness; meanwhile, an alumina ball with 20 N was used to conduct the sliding ball-on-plate friction test. The composites were chosen by varying the filler with different mass fractions. Four composites named A, B, C, and D were used with different filler mass fractions of 73, 56, 82, and 54 wt% respectively. Composite C with 82 wt% inorganic filler had shown a low COF and wear loss. In addition, the lowest wear volume for composite C after the reciprocating sliding test was nearly 0.3 $mm^3$ [62]. T. Barot et al. [63] have studied the mechanical and biological properties of novel resin-based dental composites by the addition of different silver/HNT particles as fillers. A bis-GMA/TEGDMA dental resin composite was chosen in their experiment. Subsequently, different concentrations of Ag/NHT were incorporated into the resin. The introduction of silver nanoparticles was to improve the antibacterial property of the composite, which is a current challenge to the researchers when compared to normal glass fillers. However, in their investigation, 1%–5% HNT/Ag showed better mechanical properties in comparison to 7%–10% HNT/Ag. Hence, higher weight concentration HNT/Ag showed weaker mechanical properties of resin-based composites as a dental restoration. Gan X and co-workers [64] assessed the friction/wear performance of five different resin-based dental restorative materials in three distinct surrounding media including artificial saliva, cola soft drink, and distilled water. The experiment was performed with the help of reciprocal sliding contact mode rather than pin-on-disk and other chewing simulators due to its swiftness. However, different experimental parameters with 20 N load in 0.25 Hertz frequency with reciprocating amplitude of 1000 micro-meter were chosen in their test evaluation. AP-X, Filtek P60, Z 350, VITA ZETA, and VITA LC along with various fillers accompanied with unlike filler sizes dissemination were adopted in their experiment. The role of surrounding media is to reduce the COF. The wear resistance of P60 was found to be better than those of other dental composites due to higher filler weight distribution.

# 1.5 BIOTRIBOLOGY OF OTHER HUMAN ORGANS

## 1.5.1 Cardiovascular System

The cardiovascular or circulatory system is an essential fluidic transport system of any living animal's body. On a single day, a healthy human heart beats over 100,000 times to pump blood across the body's 60,000 miles of blood vessels. The total volume of blood involved in this process is approximately 7500 liters. In an adult human body, the cardiovascular system is a highly sophisticated transport system consisting of veins, arteries, blood, and the heart itself. Such an extremely complex yet vital system may lead to the development of cardiac diseases across all living species [65].

Cardiac diseases primarily arise because of the abnormality in the function of its key organs or tissues such as the heart or the veins or arteries. Some typical examples are ischemic or coronary artery (heart) disease or CAD, heart malfunction, heart valve problems, vascular disease such as stroke, peripheral artery disease (PAD), and so on.

Together, they contribute a majority of the causes of death worldwide among noncontagious diseases. On many occasions, such a lethal disease can be treated well with the application of mechanical devices or apparatus, when other remedial options do not exist.

The massive use of coronary heart valves, pacemakers, articular prosthesis, stents, catheters, devices for heart assist, and so on is becoming popular since the last few decades. These mechanical tools come in close contact with our blood in a counter motion. Therefore, the critical design and manufacture of such devices are vital to prevent any damage of veins or arteries in contact with this viscous liquid, while simultaneously permitting adequate liquid for its proper functioning.

There are three primary mechanisms by which tribological effects are experienced by cardiac devices, based on the friction couples.

The most common type of tribological complications that arise from cardiac complications are those that involve the movement of the mechanically inserted tools that generate wear and friction.

Synthetic heart valves and mechanical circulatory support devices are some common examples.

Apart from the mechanical friction, the fluidic motion of the blood often generates friction and wear on most of the cardiovascular devices. Second, the continuous flow of the bloodstream produces fluid-friction on the surface of all kinds of cardiac devices. In addition to the above two, friction may also be generated at some point in the insertion time or as routine operation within the body, amid the devices and soft individual tissue. Artificial heart valves (mechanically operated) are very widely used devices and are extensively used to deal with aortic valve replacements [66]. Heart valve (Figure 1.9) components often experience wear and/or tear, or erosion [67]. A possible remedy to this problem is to incorporate novel coatings to curtail the development of thrombosis. Moreover, the wear resistance of the leaflets was also improved by it [68].

Similarly, vascular stents often experience friction or resistance once they are inserted [69]. The zone of friction primarily forms in the vicinity between the stent and the blood vessel [70]. Pacemakers or defibrillators in a cardiovascular system

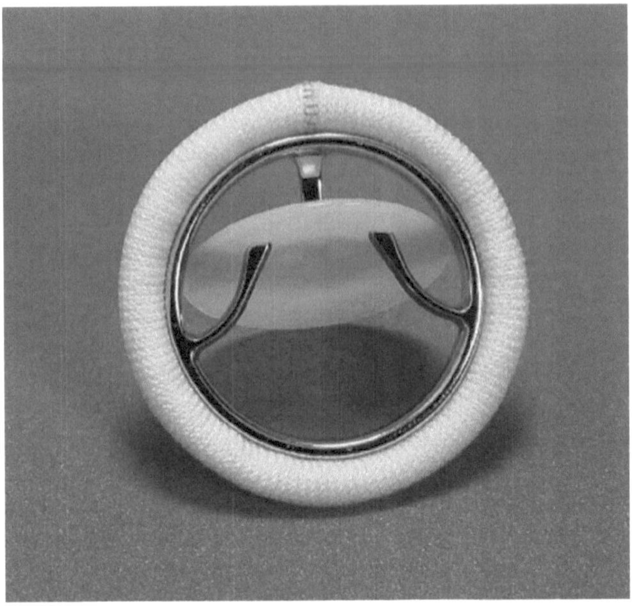

FIGURE 1.9 Tilting-disc valve. (Wikipedia.)

also experience wear and tear within the lead tissue, especially when two or more combination materials are in use [71].

### 1.5.2 Hair

Hair is an organic material, mostly protein-based, which forms in a filament-like conformation and originates from hair follicles underneath the skin surface.

In contrast to artificial fibers, which possess an identical diameter, the outer cell line thickness of human hair fibers varies within 0.5–1.0 mm thickness and 40–100 mm length. The architecture is similar as tiles are distributed on a roof in a rural house. The usual diameter of hair ranges between 50 and 100 μm.

With the variation of the sliding direction, the frictional properties of dry hair substantially differ along its fiber length. In contrast, the properties of artificial fibers are largely constant.

Bowen et al. [72] presented the development of the lateral load in correlation with the number of cycles for two mutually perpendicular treated hair fibers, experiencing a relative sliding motion whereas a normal compressive load of 10 mN is retained. In their research, sliding at a velocity of 0.4 mm/s acting in a reciprocating way, across a path of 0.8 mm, was achieved at room temperature and ~50% relative humidity.

### 1.5.3 Ocular Systems

Winking of the human eye is a natural and spontaneous process. It, however, illustrates an exciting tribological interface among the cornea of the eye and the eyelid.

The effect becomes severe for those who wear contact lenses. These types of specially fabricated lenses are set accurately on the eye and adjust its curving to furnish vision improvement. Ocular tribology is primarily concerned with the mechanisms of contact lens lubrication.

Therefore, the development of superior quality contact lens materials with characteristics that lower their friction is essential to provide comfort to the end users.

Today, nearly 140 million people on a global scale use contact lens. Furthermore, it has been found that nearly 50% of them are suffering from dry eye syndrome especially in the evening hours [73]. Contact lens often shows irritation, uneasiness, and dryness of human eyes. Therefore, a substantial segment of the users (~25%) postpone its usage, and ~25%–30% of the total users reduce the extent of use in a frequent manner [74].

Today, soft contact lenses are typically produced using hydrogels as the primary element. Hydrogels are three-dimensional (3D) hydrophilic polymers that absorb a large amount of water while preserving their structure owing to physicochemical cross-linking of various polymeric groups.

At present, various SiHy (silicone hydrogel)-based lenses commercially utilize strategies that try to enhance lubrication and improve the in-eye ease.

AcuvueOasyss (senofilcon A) is one such commercially available material synthesized from a high MW polyvinylpyrrolidone (PVP) monomer. To decrease the in-vivo lens friction, which eventually leads to the eye comfort of the users, is to use naturally occurring lubricants that are available at the ophthalmic surface.

A mucin-like glycoprotein called Proteoglycan 4, or lubricin, is found within the planes of ocular tissues [75]. In in-vitro experiments, this gel-like substance has shown its ability to reduce friction in both human cornea–polydimethylsiloxane (PDMS) and human cornea–eyelid tribopairs [75,76].

Researchers have tried to establish the performance and long-term potential of ocular lubricants as well as exceptional lens materials. To do so, a procedure for the evaluation of friction between lens and lubricants is essential.

A computational fluid mechanics study using contact lens has calculated that the eyelid applies pressures in the range of 12–18 kPa (sliding velocities of 10–100 mm/s) [77]. Morrison et al. [76] and Schmidt et al. [75] proposed a method that utilizes an in-vitro setup that has been used to successfully determine the frictional properties of materials for the contact lens and optical surfaces.

## 1.6 BIOTRIBOLOGY IN NATURE

In the field of biotribology, researchers and scientists are extensively putting their effort to explore the functionality of tribological properties of various animal and plant species. Biomimetics or bio-inspired science, although being in its infancy, has shown prospects for innovation and applications in various technological fields.

For example, the extreme hydrophobic properties of gecko skin offering anti-wetting and self-cleaning features have the potential in advanced engineering applications.

Such characteristics can also be found in tree frogs and few insects. Similarly, shark skins with dedicated nanoparticle layers have the ability to reduce surface frictions and provide a unique part of surface texturing.

### 1.6.1 Animal Feet and Surfaces

Nature exhibits various exciting tribological phenomena in various forms. Insects, reptiles, or other animals show various behavioral aspects based on their tribological characteristics in everyday life. In the animal kingdom, the inherent structure of the specific organ surfaces dealing with friction and adhesion offers significant contributions to their mechanical properties. However, in practice, this aspect has often been ignored, until recently, the field of biomimetics came into existence.

The first and most notable example is a gecko's feet, which exhibits exceptional adhesion properties and has attracted the scientific community for the past few decades. They offer tremendous potential in several biotribological applications in various mechanical systems [78,79].

It has been seen that the dorsal and ventral regions of an adult gecko's foot comprise approximately half a million spinules (hairs) or setae [79]. The typical length of these spinules spanned between hundreds of nanometers and few microns in length.

All that adds in converting gecko as the largest species in nature, which produces its superior adhesion force while fulfilling the ultimate purpose of wild movement on any surface.

Apart from its feet, the unusual surface structure of their skin makes them superhydrophobic in nature, preventing gecko's surface layer from anti-wetting properties, keeping the self-clean nature by moving the nanometric water drops at slower velocities.

Another such nature-inspired application lies in the field of soft robotics to utilize operational forces in unconventional environments, which is mimicked by special characteristics of an *Octopus vulgaris* [80]. This fascinating three-dimensional movement of the octopus, comprising crawling, gripping, and the ability to deploy its identical limb, offers an opportunity to implement in the ingeniously fabricated multifaceted soft robotic structure that has the ability to execute various tasks. Barnes et al. [81] described the adhesion behavior of a typical tree frog r in both adults and juveniles. They used white light interferometry and micro-indentation to investigate the 3-D surface structure and physicomechanical behavior of toe pads in tree frogs (*Litoria caerulea*).

Their analysis concludes that this tree frog possesses incredibly soft toe pads among all classes of biological species and it only resembles jellyfish mesogloea. It is interesting to observe that its outer surface layer is stiffer than the material underneath. Such structural assembly conforming with a softer inner structure and hard outer layer eventually leads to higher and easier adhesion of the pad with its substrate. Adult tree frogs enjoy lower adhesion values compared to the juveniles and it is believed to be one of the reasons for improved success for the elderly toe pads in many animal species.

## 1.7 SUMMARY AND PROSPECTS

The science and technology of two mutually exclusive planes that are subjected to relative motion is known as tribology and this field of science involves the study and analysis of friction, wear, and lubrication.

Biotribology, on the other hand, is a relatively recent development that has emerged from the classical field of tribology and it encompasses the tribological processes in natural organs and at the interfaces between living and implanted artificial surfaces. This fascinating field is directly associated with biological sciences, surface mechanics, physics and chemistry of materials in addition to mechanical and design engineering. Within the living species, it associates the regular function of various tissues, comprising articular cartilage, fluid vessels within the body, heart, ligaments, skin, and so on.

The advent in the field of biotribology can be seen in the applications, for example, synovial joints and artificial replacements; various usage of screws and plates in the healing of fractured bones, erosion or gradual loss of denture and restorative ingredients, skin-related friction, wear of artificial heart valves, biotribology of ocular materials such as contact lenses, and numerous other examples of biotribology at the micro- and nanoscale.

For instance, osteoarthritis is an extremely common problem in the present century, especially for the elderly people. To be precise, synovial joints related to the hip and knee joints or spines are mostly susceptible to this difficulty. Based on the widespread threat it possesses, a comprehensive investigation of implant behavior from a biotribological viewpoint is becoming extremely important and creates a wide opportunity to the scientific community.

In a nutshell, it can be concluded that the knowledge of adhesion, lubrication, and wear in the field of biotribology, not only in humans but also in various other animal species, can successfully lead to technological innovation which further helps us to develop biocompatible and functionally superior biomaterials for human benefits.

## ACKNOWLEDGMENT

The author Dr. S K Sinha expresses his gratitude to Dr. Saurav Kumar for his constant encouragement in writing up this chapter.

## CONFLICT OF INTEREST

The authors confirm that this chapter content has no conflict of interest.

## REFERENCES

1. Ian M. Hutchings, Leonardo da Vinci's studies of friction, *Wear* 360(Supplement C), 51–66, (2016).
2. D. Dowson, V. Wright, Bio-tribology, in: Proceeding of the Conference on the Rheology of Lubrication, The Institute of Petroleum, The Institution of Mechanical Engineers, and the British Society of Rheology, London, pp. 81–88, (1973).
3. R. Sonntag, J. Reinders, J.S. Rieger, D.W.W. Heitzmann, J.P. Kretzer, Hard-on-hard lubrication in the artificial hip under dynamic loading conditions, *PLoS One* 8(8), e71622, (2013).
4. J. Charnley, The lubrication of animal joints in relation to surgical reconstruction by arthroplasty, *Ann Rheum Dis*, 19(1), 10–19, (1960)

5. A.M. Trunfio-Sfarghiu, Y. Berthier, M.-H. Meurisse, J.-P. Rieu, Multiscale analysis of the tribological role of the molecular assemblies of synovial fluid. Case of a healthy joint and implants, *Tribo Int* 40(10–12), 1500–1515, (2007).
6. V.C. Mow, W.C. Hayes, *Basic Orthopaedic Biomechanics Hardcover*, Raven Press, New York, (1997).
7. J. Charnley, *Low Friction Anthroplasty of Hip: Theory and Practice*, Springer, New York, (1979).
8. G.K. McKee, J. Watson-Farrar, Replacement of arthritic hips by the McKee-Farrar prosthesis, *J Bone Joint Surg Br* 48-B (2), 245–259, (1966).
9. L.L. Verde, D.Fenga, M. Silvia Spinelli, F. Rosario Campo, M. Florio, M.A. Rosa, Catastrophic metallosis after tumoral knee prosthesis failure: A case report, *Int J Surg Case Rep* 30, 9–12, (2017).
10. T.M. Grupp, B. Fritz, I. Kutzner, C. Schilling, G. Bergmann, J. Schwiesau, Vitamin E stabilized polyethylene for total knee arthroplasty evaluated under highly demanding activities wear simulation, *Acta Biomater* 48, 415–422, (2017).
11. L. Lapaj, J. Wendland, J. Markuszewski, A. Mroz, T. Wisniewski, Retrieval analysis of titanium nitride (TiN) coated prosthetic femoral heads articulating with polyethylene, *J Mech Behav Biomed Mater* 55, 127–139, (2015).
12. F. Beyer, C. Lutzner, S. Kirschner, J. Lutzner, Midterm results after coated and uncoated TKA: A randomized controlled study. *Orthopedics* 39(3 Suppl), S13–S17, (2016).
13. A.L. Puente Reyna, B. Fritz, J. Schwiesau, C. Schilling, B. Summer, P. Thomas, T.M. Grupp, Metal ion release barrier function and biotribological evaluation of a zirconium nitride multilayer coated knee implant under highly demanding activities wear simulation, *J Biomech* 79, 88–96, (2018).
14. P. Agache, P. Humbert, *Measuring the Skin*, Springer-Verlag, ISBN: 978-3-642-05691-8, XXIII, 784, (2004).
15. M.F. Leyva-Mendivil, J. Lengiewicz, A. Page, N.W. Bressloff, G. Limbert, Skin microstructure is a key contributor to its friction behaviour, *Tribol Lett* 65, 1–17, (2017).
16. K.A. Holbrook, G.F. Odland, Regional differences in the thickness (cell layers) of the human stratum corneum: An ultrastructural analysis, *J Invest Dermatol* 62, 415, (1974)
17. B.N.J. Persson, A. Kovalev, S.N. Gorb, Contact mechanics and friction on dry and wet human skin, *Tribol Lett* 50, 17–30, (2013).
18. D. Fawcett, *Text Book of Histology*, W.B. Saunders Co., Philadelphia, (1986).
19. L. Vilhena, A. Ramalho, Lubricants: Friction of human skin against different fabrics for medical use, *Lubricants* 4, 6, (2016).
20. L.C. Gerhardt, V. Strässle, A. Lenz, N.D. Spencer, S.N. Derler, Influence of epidermal hydration on the friction of human skin against textiles, *J R Soc Interface* 5, 1317–1328, (2008).
21. M. Adams, B. Briscoe, S. Johnson, Friction and lubrication of human skin, *Tribol Lett* 26 (3), 239–253, (2007).
22. L.J. Wolfram, Frictional properties of skin, in: *Cutaneous Investigation in Healthy and Disease Noninvasive Methods and Instrumentation*, J.L. Leveque (Ed.), MaracelDefier, New York, pp. 49–57, (1989).
23. A.F. El-Shimi, In vivo skin friction measurements, *J Soc Cosmet Chem* 28, 37–51, (1977).
24. J.S. Comaish, E. Bottoms, The skin and friction: Deviations from amonoton's laws, and the effects of hydration and lubrication, *Br J Dermatol* 84, 37–43, (1971).
25. H. Zahouani, G. Boyer, C. Pailler-Mattei, M. Ben Tkaya, R. Vargiolu, Effect of human ageing on skin rheology and tribology, *Wear* 271, 2364–2369, (2011).
26. N.K. Veijgen, M.A. Masen, H.E. Van der Heide, Relating friction on the human skin to the hydration and temperature of the skin, *Tribol Lett* 49, 251–262, (2013).

27. N. Veijgen, Skin friction- A novel approach to measuring In-vivo human skin, PhD Thesis, University of Twente, (1913).
28. W. Tang, B. Bhushan, Adhesion, friction and wear characterization of skin and skin cream using atomic force microscope, *Colloids Surf B: Biointerfaces* 76, 1–15, (2010).
29. S. Derler, L.C. Gerhardt, Tribology of skin: Review and analysis of experimental results for the friction coefficient of human skin, *Tribol Lett* 45:1–27, (2012).
30. S. Derler, G.M. Rotaru, W. Ke, L. El Issawi-Frischknecht, P. Kellenberger, A. Scheel-Sailer, R.M. Rossi, Microscopic contact area and friction between medical textiles and skin, *J Mech Behav Biomed Mater* 38, 114–125, (2014).
31. S. Derler, U. Schrade, L.C. Gerhardt, Tribology of human skin and mechanical skin equivalents in contact with textiles, *Wear* 263(7–12), 1112–1116, (2007).
32. W. Tang, S.R. Ge, H. Zhu, X.C. Cao, N. Li, The influence of normal load and sliding speed on frictional properties of skin, *J Bionic Eng* 5, 33–38, (2014).
33. M. Zhang, A.F. Mak, In vivo friction properties of human skin, *Prosthet Orthot Int* 23, 135–141, (1999).
34. L. Meredith, J. Brown, E. Clarke, Relationship between skin abrasion injuries and clothing material characteristics in motorcycle crashes, *Biotribology* 3, 20–26, (2015).
35. Z.R. Zhou, Z.M. Jin, Biotribology: Recent progresses and future perspectives, *Biosurf Biotribol* 1(1), 3–24, (2015).
36. L.H. Mair, Wear in dentistry—Current terminology. *J Dent* 20(3), 140–144, (1992).
37. E. Neyraud, M. Morzel, Biological films adhering to the oral soft tissues: Structure, composition, and potential impact on taste perception, *J Texture Stud* 50, 19–26, (2019).
38. F. Xu, L. Laguna, A. Sarkar, Aging-related changes in quantity and quality of saliva: Where do we stand in our understanding? *J Texture Stud*, 50, 27–35, (2019).
39. Z.R. Zhou, J. Zheng, Tribology of dental materials: A review, *J Phys D: Appl Phys*, 41(-11), 113001, (2008).
40. Z.R. Zhou et al., *Dental Biotribology*, © Springer Science+Business Media, New York, (2013).
41. L.H. Mair, T.A. Stolarski, R.W. Vowles, C.H. Lloyd, Wear: Mechanisms, manifestations and measurement. Report of a workshop, *J Dent* 24, 141–148, (1996)
42. R.G. Every, W.G. Kuhne, Biomodal wear of mammalian teeth, in: D.M. Kermack, K.A. Kermack (Eds.), *Early Mammals*, Academic Press, London, pp. 23–27, (1971).
43. R. Berlin, L. Dessner, Bruxism and Chronic headache, *The Lancet* 276(7145), 289–291, (1960).
44. H. Thomson, Functions of the masticatory system, in: H. Thomson (Ed.), *Occlusion*. 2nd ed., John Wright and Sons, Bristol, pp. 87–97, (1975).
45. M.C.D.N.J.M. Huysmans, H.P. Chew, R.P. Ellwood, Clinical studies of dental erosion and erosive wear, *Caries Res*, 45(s1), 60–68, (2011).
46. T. Molleson, K. Jones, Dental evidence for dietary change at Abu Hureyra, *J Archaeol Sci* 18, 525–539, (1991).
47. B.G.N. Smith, N.D. Robb, Dental erosion in patients with chronic alcoholism, *J Dent* 17(5), 219–221, (1989).
48. T. Jaeggi, A. Lussi, Toothbrush abrasion of erosively altered enamel after intraoral exposure to saliva: An in-situ study, *Carious Res* 33, 455–461, (1999).
49. F. Rupp, L. Liang, J. Geis-Gerstorfer, L. Scheideler, F. Hüttig, Surface characteristics of dental implants: A review, *Dent Mater* 34(1), 40–57, (2018).
50. F.J. Gil, E Fernández, J.M. Manero, J.A. Planell, J. Sabrià, M. Cortada, L. Giner, A study of the abrasive resistance of metal alloys with applications in dental prosthetic fixators, *Biomed Mater Eng* 212, 161–167, (1995).
51. D. Iijima, T. Yoneyama, H. Doi, H. Hamanaka, N. Kurosaki, Wear properties of Ti and Ti–6Al–7Nb castings for dental prostheses, *Biomaterials* 24, 1519–1524, (2003).

52. C. Ohkubo, Wear resistance of experimental Ti–Cu alloys, *Biomaterials* 24(20), 3377–3381, (2003).
53. D.J. Todd, D. Burrows, Nickel allergy in relationship to previous oral and cutaneous nickel contact, *Ulster Med J* 58:168–171, (1989).
54. H. Doi, T. Yoneyama, E. Kobayashi, & T. Hanawa, Fatigue property of Ti-5Al-13Ta alloy dental castings in 0.9% NaCl solution, *Mat Trans* 47(9), 2444–2447, (2006).
55. G. Rahilly, N. Price, Current products and practice, *J Orthod* 30(2), 171–174, (2003).
56. D. Salamon, Chapter 6-Advanced ceramics, in: J.Z. Shen, & T. Kosmač (Ed), *Advanced Ceramics for Dentistry*, Elsevier, pp. 103–122, (2014).
57. M. Hisbergues, S. Vendeville, P. Vendeville, Zirconia: Established facts and perspectives for a biomaterial in dental, *J Biomed Mater Res B Appl Biomater* 88:519–529, (2009).
58. A. Ioannidis, et al. Load-bearing capacity of CAD/CAM 3D-printed zirconia, CAD/CAM milled zirconia, and heat-pressed lithium disilicate ultra-thin occlusal veneers on molars, *Dent Mater* 36(4):e109–e116, (2020).
59. H.Y. Yu, Z.B. Cai, P.D. Ren, M.H. Zhu, & Z.R. Zhou, Friction and wear behavior of dental feldspathic porcelain, *Wear* 261(5–6), 611–621, (2006).
60. R. Talib, Dental composites: A review, *J Nihon Univ Sch Dent* 35(3), 161–170, (1993).
61. X Hu, Q Zhang, J Ning, W Wu, C Li. Study of two-body wear performance of dental materials, *J Natl Med Assoc* 110(3), 250–255, (2017).
62. J.C.M. Souza, A.C. Bentes, K. Reis, S. Gavinha, M. Buciumeanu, B. Henriques, J.R. Gomes, Abrasive and sliding wear of resin composites for dental restorations, *Tribol Int* 102, 154–160, (2016).
63. T. Barot, D. Rawtani, P. Kulkarni, Physicochemical and biological assessment of silver nanoparticles immobilized Halloysite nanotubes-based resin composite for dental applications, *Heliyon* 6(3), e03601, (2020).
64. X. Gan, Z. Cai, B. Zhang, X Zhou, H. Yu, Friction and wear behaviors of indirect dental restorative composites, *Tribol Lett* 46(1), 75–86, (2012).
65. L.H. Young, Heart disease in the elderly, in: B.L. Zaret, M. Moser, & L.S. Cohen (Eds.), *Yale University School of Medicine Heart Book*, Yale University School of Medicine, New York, pp. 263–272, (1992).
66. D.F. Zhao, M. Seco, J.J. Wu, J.B. Edelman, M.K. Wilson, M.P. Vallely, M.J. Byrom, P.G. Bannon, Mechanical versus bioprosthetic aortic valve replacement in middle-aged adults: A systematic review and meta-analysis, *Ann Thorac Surg* 102(1) 315–327, (2016).
67. S. Aoyagi, S. Fukunaga, K. Arinaga, Disc wear and entrapment in a Starr-Edwards mitral caged-disc valve, *J Heart Valve Dis* 20(4), 474–476, (2011).
68. M.J. Jackson, G.M. Robinson, N. Ali, Y. Kousar, S. Mei, J. Gracio, H. Taylor, W. Ahmed, Surface engineering of artificial heart valve disks using nanostructured thin films deposited by chemical vapour deposition and sol-gel methods, *J Med Eng Technol* 30(5), 323–329, (2006).
69. A. Lelasi, A. Anzuini, Guide-catheter extension system facilitated multiple bioresorbable vascular scaffolds (ABSORBs) delivery in a very long and resistant coronary artery lesion), *Cardiovasc Revasc Med* 15(2), 117–120, (2014).
70. A. Prasad, N. Xiao, X.Y. Gong, C.K. Zarins, C.A. Figueroa, A computational framework for investigating the positional stability of aortic endografts, *Biomech Model Mechanobiol* 12(5), 869–887, (2013).
71. B. Małecka, A. Ząbek, A. Ciaś, J. Stępiński, A. Kutarski, J. Rońda, J. Lelakowski, J. Małecki, Endocardial silicone lead wear: Description of tribological phenomena on the basis of microscopic examination of removed leads. Preliminary report, *Kardiol Pol* 72(10), 960–968, (2014).

72. J. Bowen, S.A. Johnson, A.R. Avery, M.J. Adams, Friction and wear of human hair fibres, *Surf Topogr: Metrol Prop* 4(2), 024008, (2016).
73. C. Riley, G Young, R. Chalmers Prevalence of ocular surface symptoms, signs, and uncomfortable hours of wear in contact lens wearers: The effect of refitting with daily-wear silicone hydrogel lenses (senofilcon a), *Eye Contact Lens* 32(6), 281–286, (2006).
74. M.A. Lemp, The definition and classification of dry eye disease: report of the Definition and Classification Subcommittee of the International Dry Eye WorkShop (2007), *Ocul Surf* 5(2), 75–92, (2007).
75. T.A. Schmidt, D.A. Sullivan, S.M. Richards, N. Knop, S. Liu, A. Sahin, R.R. Darabad, S. Morrison, W.R. Kam, B.D. Sullivan, Transcription, translation, and function of lubricin, a boundary lubricant, at the Ocular surface, *JAMA Ophthalmol* 131(6), 766–776, (2013).
76. S. Morrison, D.A. Sullivan, B.D. Sullivan, H. Sheardown, T.A. Schmidt, Dose- dependent and synergistic effects of proteoglycan 4 on boundary lubrication at a humancornea–polydimethylsiloxane biointerface, *Eye Contact Lens* 38(1), 27–35, (2012).
77. A.C. Dunn, J.A. Tichy, J.M. Urueña, W.G. Sawyer, Lubrication regimes in contact lens-wear during a blink, *Tribol Int* 63, 45–50, (2013).
78. A. Anand, M.I.U. Haq, A. Raina, K. Vohra, R. Kumar, S.M. Sharma, Natural Systems and TribologyAnalogies and Lessons, *Mater Today: Proc* 4(4), 5228–5232, (2017).
79. K. Autumn, Y. A. Liang, S. T. Hsieh, W. Zesch, W.P. Chen, T.W. Kenny, R. Fearing, R.J. Full, Adhesive force of a single gecko foot-hair, *Nature* 405(6787), 681–685, (2000).
80. M. Calisti, M. Giorelli, G. Levy, B. Mazzolai, B. Hochner, C. Laschi, P. Dario, An octopus-bioinspired solution to movement and manipulation for soft robots, *Bioinspir Biomim* 6(3), 036002, (2011).
81. W.J.P. Barnes, P. Perez-Goodwyn, S.N. Gorb, Mechanical properties of the toe pads of the treefrog, Litoriacaerulea, *Comp. Biochem. Physi. A* 141(3), S145, (2005).

# 2 Nano Lubricant Additives

*Tianyi Sui*
Tianjin University

## CONTENTS



## 2.1 INTRODUCTION

In recent years, there has been growing interest in investigating nanomaterials, mainly in size between 1 and 500 nm, as lubricant additives [1]. With the improvement of nanotechnology, advanced nanomaterials are synthesized and have special properties in the tribology field. The word "nanotechnology" was firstly raised in the famous speech "There's Plenty of Room at the Bottom" by the physicist Richard Feynman on December 29, 1959. In 1968, Werner Stöber invented the Stöber method and realized the size control of silica nanoparticles [2]. Harold W. Kroto, Robert F. Curl Jr., and Richard E. Smalley discovered fullerene in 1985 and won the Nobel Prize in 1996. With the progress of nanotechnology, many researchers all over the world synthesized different kinds of nanomaterials. Those nanomaterials could be divided into four types: zero-dimension (e.g., silver and gold colloids), one-dimension (e.g., carbon nanotubes and silver nanowires), two-dimension (e.g., graphene), and three-dimension materials (e.g., nanoparticles). Parameters such as structure, size, and surface polarity would influence their properties significantly. Figure 2.1 shows

DOI: 10.1201/9781003096443-2

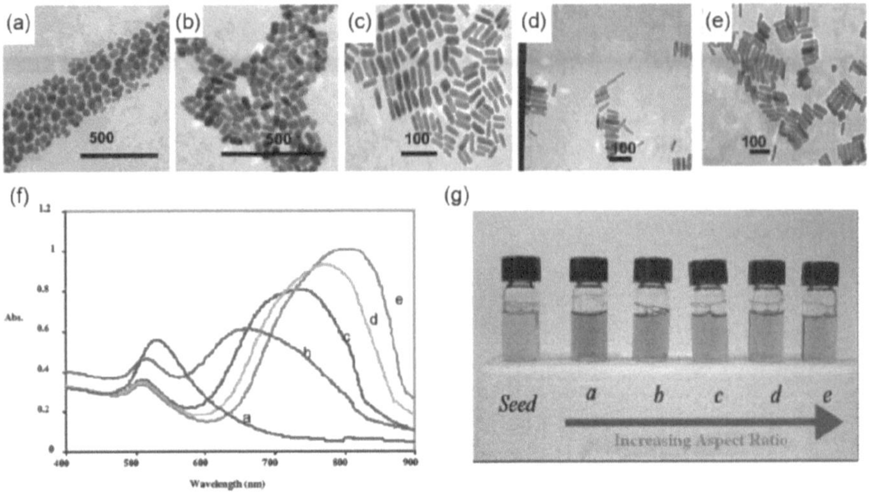


**FIGURE 2.1** Transmission electron micrographs (a–e), optical spectra (f), and photographs of (g) aqueous solutions of gold nanorods of various aspect ratios. (Images are taken from Ref. [3] with permission from the American Chemical Society.)

the change of the optical spectra and color with different kinds of gold nanoparticles. Nanomaterials were broadly investigated and widely used in many fields. This section will focus on the application of nanomaterials in the tribology field as lubricant additives.

Lubricant additives are used to clean the contact surface, seal the contacting parts, remove heat from the surface, and, most importantly, reduce friction and wear. With small proportions of additives, the rheology, anti-wear, and friction reduction properties would change dramatically. With a decrease in the particle size, the proportion of atoms on the surface of nanoparticles increases rapidly. The proportion of atoms at the surface reaches 20% when the particles' size decreases to 10 nm; almost all of the atoms appear on the surface when the particle size is 1 nm. This phenomenon leads to notable changes in the properties of particles. In the field of tribology, the high specific surface area and the small size of the nanomaterial provide the perfect platform for improving the lubrication behavior when used as a lubricant additive. This chapter will provide a short review of the preparation, dispersion, and tribological performance of some best-known nanomaterials for lubricant additives. Moreover, the anti-wear and friction reduction mechanisms for different kinds of nanomaterials will also be discussed.

## 2.2 PREPARATION OF A NANO LUBRICANT ADDITIVE

This section introduces the preparation methods for typical nanomaterials in each category. The nanomaterials used as lubricant additives can be divided into the categories listed in Table 2.1.

**TABLE 2.1**
**Nanomaterial Additives Used in Tribology**

| Category | Nanomaterial Additives |
|---|---|
| Oxides | $SiO_2$, $TiO_2$, $Al_2O_3$, CuO, $Fe_3O_4$, $Fe_2O_3$ |
| | Composite particles: $TiO_2/SiO_2$, $ZrO_2/SiO_2$, $Al_2O_3/SiO_2$, ZnO/CuO |
| Sulfides | $MoS_2$, fullerene-like $MoS_2$, $MoS_2$ sheet, $WS_2$, fullerene-like $WS_2$, CuS, PbS |
| Metals | Cu, Ag, Pd, Pb, Mo, Ni |
| Carbon materials | Graphene, graphene oxide, carbon nanotubes, carbon spheres, carbon nanoonions, carbon nanohorns |

### 2.2.1 The Oxide Nano Lubricant Additives

The oxide nanoadditives are widely studied as lubricant additives. Two main preparation methods of oxide nanomaterials exist, i.e., chemical methods and physical methods. Typical chemical methods include the sol–gel method, microemulsion method, hydrolyzation–precipitation method, and chemical vapor deposition (CVD) method.

The sol–gel process is the most commonly used chemical method to prepare oxide nanoparticles, especially for $SiO_2$ and $TiO_2$ nanoparticles. The process involves two main steps. The first step is making monomers into a colloidal solution, which is termed "sol," the precursor; the second step is generating the particles by integrating the precursor and forming a network. As the content of particles is usually too low, the fluid needs to be removed, and the mixture always becomes gelatinous, which is termed "gel." During this wet-chemical process, two main reactions take place: the hydrolysis reaction described in Equation (2.1) and the polymerization reaction described in Equations (2.2) and (2.3):

$$M(OR)n + H_2O \rightarrow M(OH)x(OR)n - x + xROH \quad (2.1)$$

$$-M - OH + HO - M- \rightarrow -M - O - M - +H_2O \quad (2.2)$$

$$-M - OR + HO - M- \rightarrow -M - O - M - +ROH \quad (2.3)$$

The Stöber process is one of the best-known sol–gel processes and the most widely used synthesis method for preparing silica nanoparticles. The process was developed in 1968 by Werner Stöber and his colleagues [2]. Tetraethyl orthosilicate (TEOS) was used as the precursor and reacted with water in an alcohol solvent. The molecules then form a large cluster and become silica nanoparticles during the condensation. Figure 2.2 shows a schematic representation of the Stöber process and scanning electron microscopy (SEM) images of silica nanoparticles. Monodispersed silica nanoparticles with good spherical shapes are obtained after the condensation process. The size of nanoparticles could be adjusted by tuning the amount of ammonium hydroxide added to ethanol, and the diameter of silica nanoparticles could be controlled from 20 to 2000 nm.

The ball-milling method is one of the most commonly used physical methods to form nanooxide materials such as $SiO_2$, $TiO_2$, and $Al_2O_3$. Oxide materials are

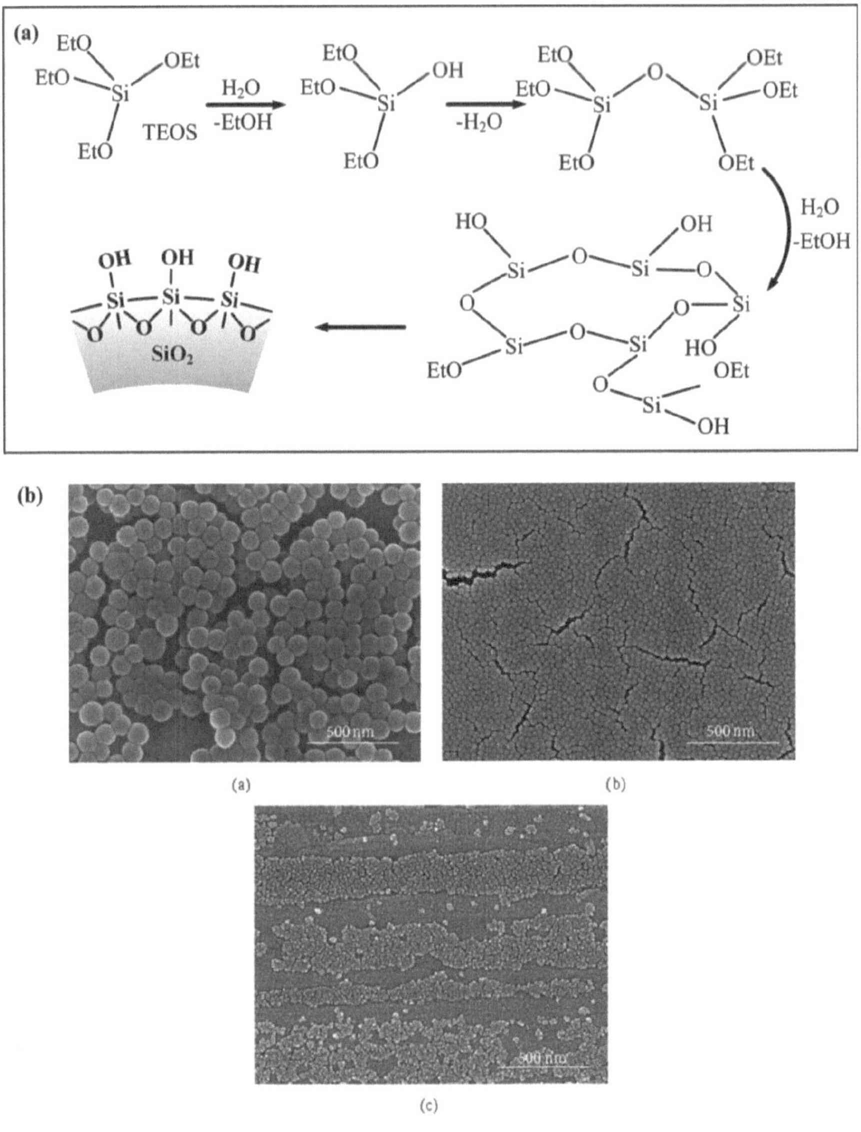


**FIGURE 2.2** (a) Schematic representation of the Stöber method. (b) Silica nanoparticles with different sizes. (From Ref. [4] with the permission of Hindawi.)

crushed and ground by a ball mill grinder. The impact, shear, and friction forces of the grinder grind the material down to the nanorange, and nanomaterials with different size ranges are then screened out. The rotating speed and the mass ratio of the mill ball and the material may affect the nanomaterials. New techniques such as ultrasonic vibration are also applied to improve the quality of the nanomaterials. The shape of the nanoparticles produced by the ball-milling method is irregular, the particle size distribution is wide, and the uniformity is poor. However, this method is economical and suitable for a large-scale preparation of nanomaterials.

### 2.2.2 The Sulfide Nano Lubricant Additives

The layered structure of sulfides such as $MoS_2$ and $WS_2$ leads to good lubrication performance when used as lubricant additives. Sulfide nano lubricant additives $MoS_2$ and $WS_2$ are the most commonly used sulfide nanomaterials in the tribology field. Figure 2.3 shows the structure and transmission electron microscopy (TEM) images of $MoS_2$ and $WS_2$. The preparation method of $MoS_2$ and $WS_2$ can be summarized as a physical and chemical method. Mechanical milling methods, especially ball-milling, are the main physical methods to prepare sulfide nanomaterials. The mechanical exfoliation method can also be used to prepare multilayer sulfide nanomaterials. Those processing methods are simple with low processing costs, but nanomaterial parameters are hard to control. Thus, chemical methods are more commonly used in synthesizing sulfide nanomaterials.

For chemical processing methods, two main methods exist for synthesizing $WS_2$ and $MoS_2$:

1. Direct reaction of a tungsten or molybdenum source with a sulfur source; the tungsten or molybdenum source can be an elementary substance or oxide. The tungsten or molybdenum source can be sulfurated by hydrogen sulfide gas according to the reactions described in Equations (2.4) and (2.5), or sulfurated by sulfur gas at high temperature described in Equation (2.6). This method has simple operation, low production costs, and a short production circle. However, both $H_2S$ and sulfur gas are harmful to human health, so special attention should be paid when handling them.

$$MoO_3 + 2H_2S + H_2 = MoS_2 + 3H_2O \tag{2.4}$$

$$MoO_3 + 3H_2S = MoS_2 + 3H_2O + 1/8S_8 \tag{2.5}$$

$$MoO_3 + 7/8S_8 = 2MoS_2 + 3SO_2 \tag{2.6}$$

2. The second method is the precursor decomposition method. The precursor is prepared using the reaction between the tungsten or molybdenum source and the sulfur source, then using the reduction or decomposition method sulfide nanomaterials are obtained. Equations (2.7) and (2.8) describe the typical decomposition method for preparing $MoS_2$. Upon heating, the $(NH_4)_2MoS_4$ decomposes and generates $MoS_3$; when the temperature is higher than 673 K, $MoS_3$ decomposes to $MoS_2$ and the sulfur element. By tuning the heating rate, the size of nanoparticles and the specific surface area can be controlled. The precursor decomposition method is simple, and the nano-$MoS_2$ shows good crystallization. However, reaction conditions such as the heating rate, end temperature, and flow rate influence the morphology significantly, making the control of morphology challenging.

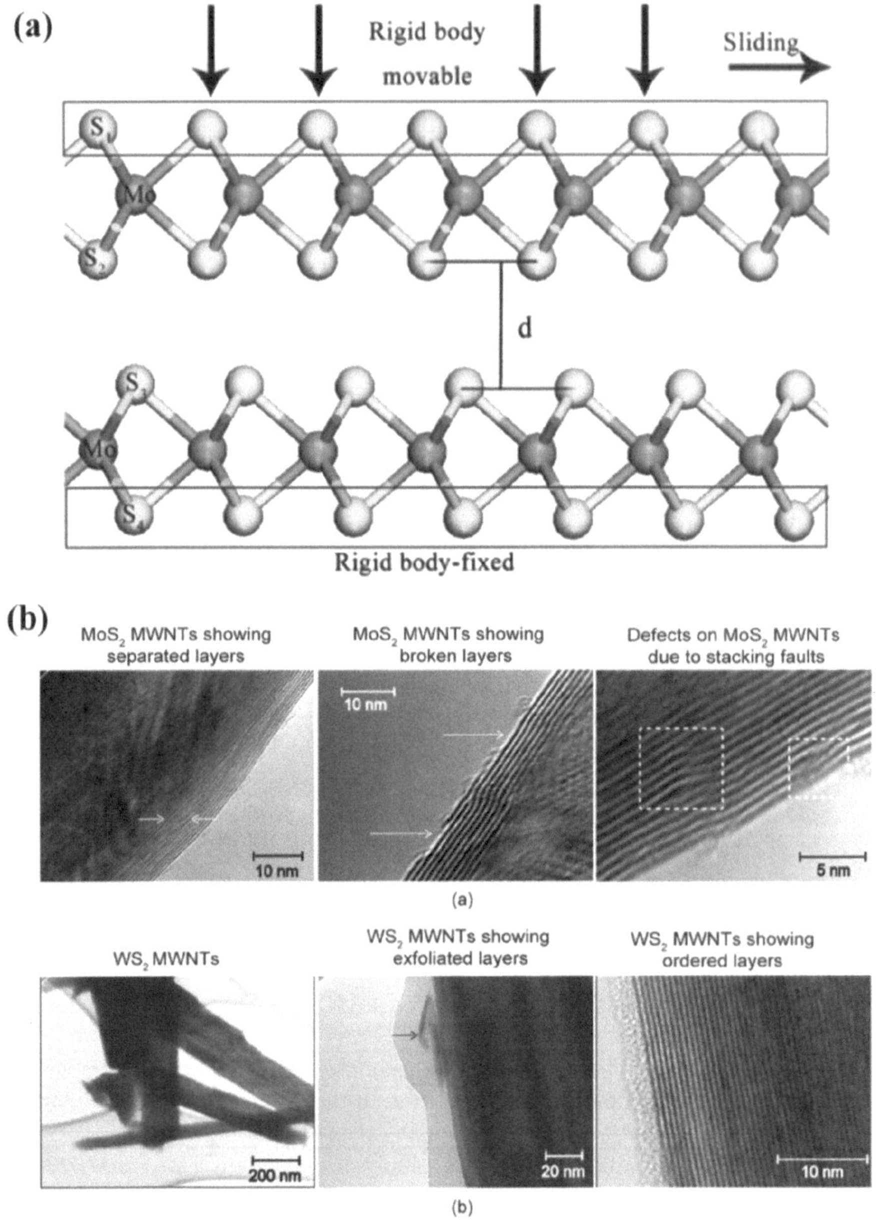


**FIGURE 2.3** (a) Schematic diagram of the $MoS_2$ crystal structure. (Image from Ref. [5] with the permission of Elsevier.) (b) TEM images of $MoS_2$ and $WS_2$. (From Ref. [6] with the permission of Springer Nature.)

$$(NH_4)_2 MoS_4 = MoS_3 + 2NH_3 + H_2S \tag{2.7}$$

$$MoS_3 = MoS_2 + 1/8 S_8 \tag{2.8}$$

### 2.2.3 The Metal Nano Lubricant Additives

The processing method for metal nano lubricant additives can be divided into two main categories: top-down and bottom-up. For the top-down method, metal materials are usually processed by physical methods, such as vacuum condensation, magnetron sputtering, and laser ablation. The physical method applies different forms of energy to metal materials (such as heat, pressure, and laser) to break the structure of the material, thus reducing the particle size from the macroscale to the nanoscale. The physical method is simple, and nanomaterials produced are of high purity, but it requires complex equipment that is of high costs.

The chemical processing method for metal nano lubricants is the bottom-up method to prepare nanomaterials. Nanomaterials are produced by controlling the growth of materials at molecular and atomic levels. The most commonly used chemical methods to prepare metal nanomaterials are the microemulsion method, the reduction method, and the electrochemical method. Figure 2.4 shows the Ag and Pb nanoparticles prepared using the chemical method.

### 2.2.4 The Carbon Nano Lubricant Additives

During the last two decades, carbon materials have attracted attention from researchers all over the world. Figure 2.5 shows the typical preparation methods for a carbon nanomaterial. Graphene was firstly discovered by physicist professor A. Geim and K. Novoselov from the University of Manchester by using a simple mechanical exfoliation method. Graphene consists of a monolayer or multilayer of carbon atoms in the form of a two-dimensional hexagonal lattice (see Figure 2.5). Until now, the mechanical exfoliation method is still one of the most used methods to prepare graphene because it is simple and leads to a high-quality product. However, it makes the control of the parameters of graphene challenging; it is an inefficient and costly method, so it is not suitable for mass production.

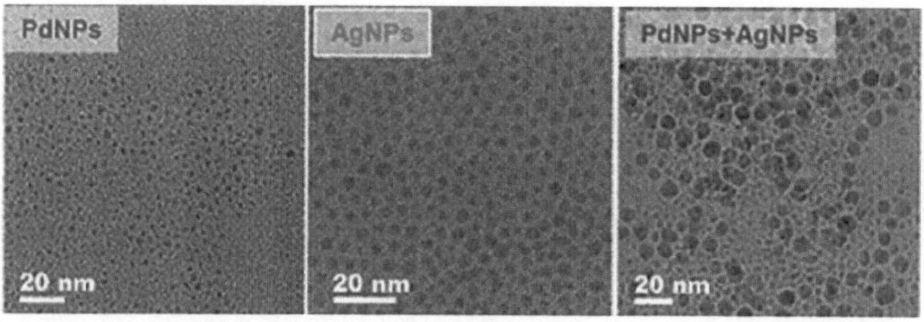


**FIGURE 2.4** The TEM morphology of Ag NPs, Pd NPs, and AgNP+PdNP mixture after eight days stored at room temperature. (Images were taken from Ref. [7] with the permission of the American Chemical Society.)

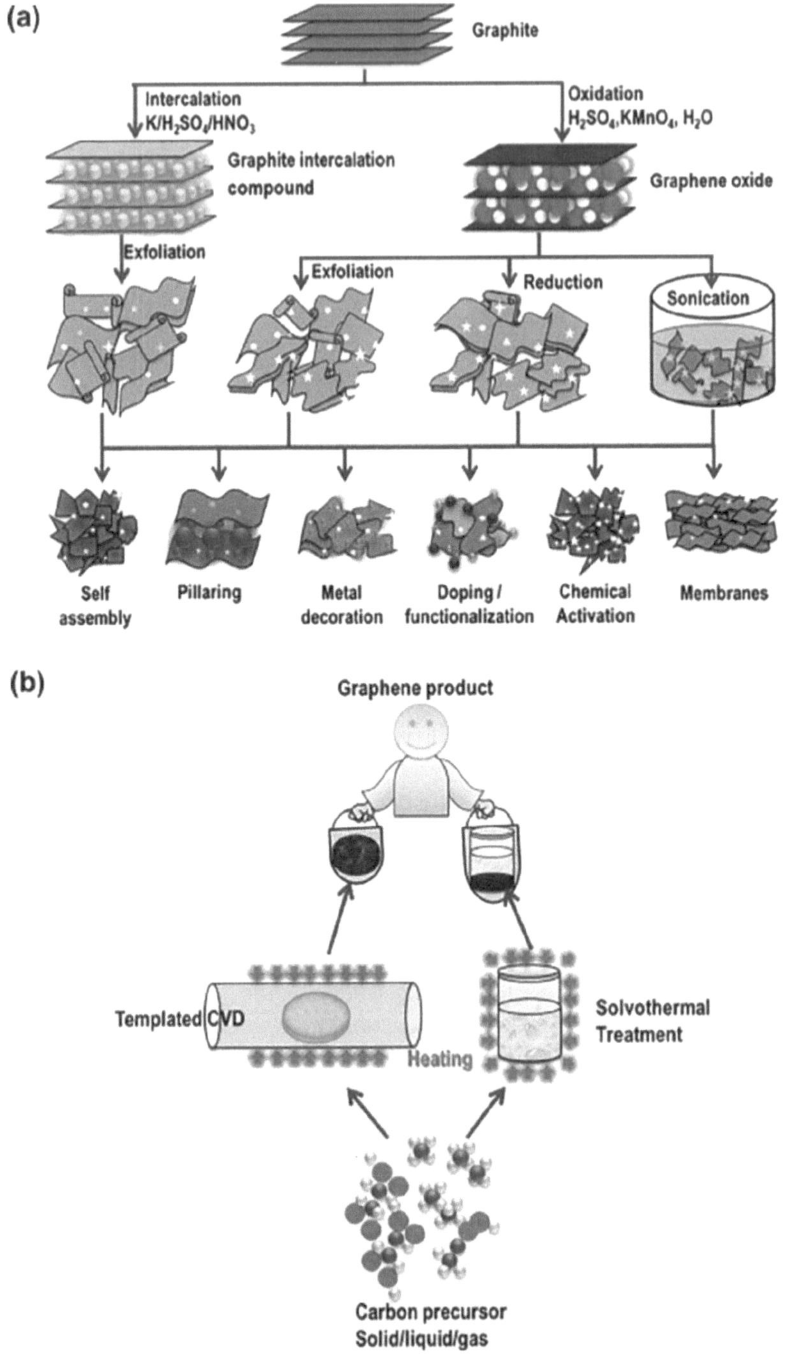


**FIGURE 2.5** (a) Graphitic top-down approach. (b) Molecular carbon precursor bottom-up approach for the massive production of a wide variety of graphene-based materials. (Images were taken from Ref. [8] with the permission of Elsevier.)

The CVD method, the graphite oxide reduction method, and the epitaxial growth method are commonly used chemical methods to prepare graphene. The CVD method is a potential method for mass production of graphene. Hydrocarbons such as methane and ethanol are injected into the surface of the metallic base Cu, and Ni heated at a high temperature, and the reaction lasted for a certain time before cooling. During the cooling process, several layers or a monolayer of graphene form on the surface of the base. The reaction can take place at low temperatures, which is useful for reducing energy consumption. This method produces graphene that separates easily by chemical etching of metal, which is suitable for later processing of graphene.

The graphite oxide reduction method is another way to prepare graphene. Graphite is oxidized to graphite oxide using a strong oxidant like concentrated sulfuric acid, concentrated nitric acid, and potassium permanganate. The spacing between graphite layers is increased by adding oxygen-containing functional groups during oxidation; graphene oxide can be easily prepared by ultrasonic processing of graphite oxide. Graphene oxide is then reduced to graphene using a strong reductant like hydrazine hydrate and sodium borohydride. This method is simple and of low-cost and allows the massive production of graphene.

## 2.3 THE INFLUENCE OF NANOADDITIVES ON THE TRIBOLOGICAL PROPERTIES OF LUBRICANTS

### 2.3.1 The Dispersity and Stability of Nanoadditives in Lubricants

The dispersity and stability of nanomaterials influence their performance as lubricant additives significantly. Nanomaterials are easy to aggregate because of their high specific surface area and high surface energy. Their aggregation usually becomes more significant when dispersing into lubricants as additives. When nanomaterials aggregate in lubricants, they will form a large cluster and harm the lubrication. The unstable dispersity of nanomaterials in lubricants is also bad for commercial application. Thus, the dispersity and stability of nanomaterials in lubricants become a bottleneck for their application in the tribology field.

However, the large surface area and high surface energy of nanomaterials provide an excellent platform for surface modification. The surface can be modified by a physical or chemical method to change the physical and chemical properties purposefully. Two main methods exist to keep nanomaterial additives well dispersed in a lubricant: changing the formation of the lubricant by adding the dispersant and modifying the nanomaterial surface with chemicals. Both methods improve the compatibility of nanomaterials and lubricants by changing the nanomaterial surface through adsorption or chemical reaction.

Adding a dispersant is the classic method to improve the dispersity of nanomaterials in lubricants. Unmodified nanomaterials such as $SiO_2$ and graphene are hydrophilic, a property that makes them hard-to-disperse in oil lubricants. Thus, dispersants like oleic acid and sorbitol monostearate are mixed with lubricant oil to improve the dispersity and stability of nanomaterials. The dispersants improve the dispersity of nanomaterials because they have amphiphilic molecules with both lipophobic and lipophilic functional groups. The lipophobic groups attach to the surface of the nanomaterials and form an organic layer that can stabilize nanoparticles in the lubricant oil.

In recent years, with the development of nanotechnology, the surface modification technique became more popular in improving the dispersity of nanomaterials. Surfactants and silane are the most widely used for surface modification. Surface modification enhances the compatibility between the nanomaterial surface and solvent, thus improving the dispersity of nanomaterials in lubricants. At the same time, chemical groups can be introduced to nanomaterial surfaces to improve the function of nanoadditives.

For surface deposition, nanomaterials are covered with inorganic substances formed by a chemical method such as the sol–gel process introduced in Section 1.2.1. Although chemical methods are used to form the cover layer of nanomaterials, the cover layer is not connected with the nanomaterial surface with a chemical bond.

Nanomaterials can also be modified by coupling agents such as silane and titanate used for modifying the surface (see Figure 2.6). The process is simple: the coupling agent reacts with the hydroxyl group on the nanomaterial surface solvent at low temperatures (usually under 100°C). The surface will be covered with the functional groups from the coupling agent; functional groups on the nanomaterial surface will change the properties of the nanomaterial. SEM images of $SiO_2$ nanoparticles (Figure 2.6) show that the dispersity of nanoparticles with different functional groups is different.

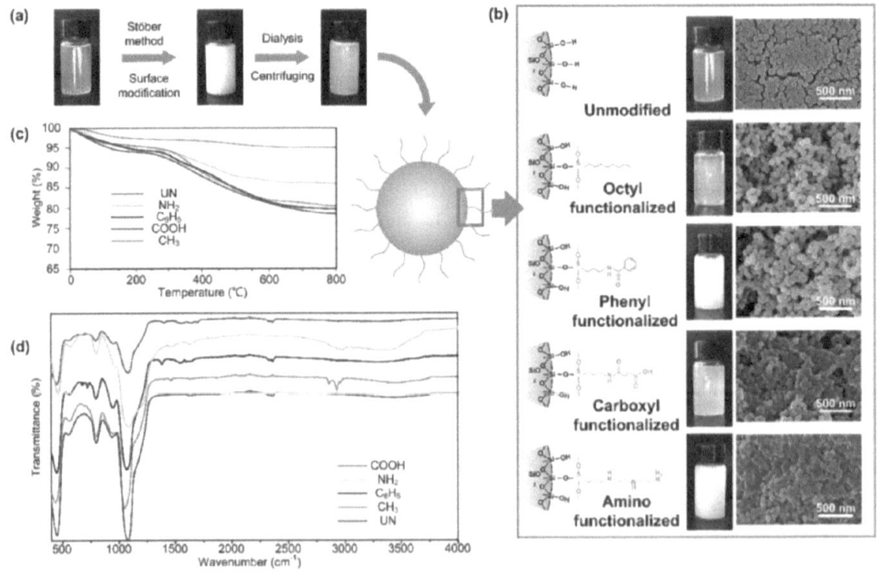


**FIGURE 2.6** The surface modification process and silica nanoparticles with different functional groups. (Images were taken from Ref. [9] with the permission of the American Chemical Society.)

The stability of a nanomaterial in the lubricant is related to many aspects, such as the lubricant viscosity, nanomaterial size, nanomaterial concentration, and compatibility between the nanomaterial and the lubricant. Usually, the stability of nanomaterials increases with the increase of lubricant viscosity and the decrease of nanomaterial size. Particles are easier to aggregate under higher concentrations. Because of the complexity of the dispersion of nanomaterials in lubricants, the stable time reported in the literature varies from several hours to a few months. As introduced in this section before, the stability of nanomaterials limits their application in the tribology field. However, nanomaterials have been successfully applied in some commercial products like nano-$TiO_2$ in Castrol EDGE engine oil.

### 2.3.2 The Rheological Properties of Lubricants with Nano Lubricant Additives

The viscosity of the lubricant is one of the most important physical properties because it influences the lubricant film thickness significantly. The film thickness is proportional to the viscosity under hydrodynamic lubrication and proportional to the 0.7 power of the viscosity of the fluid under electrohydrodynamic lubrication. As the lubricant bears part of the load, the viscosity of the lubricant influences the bearing capacity of the lubricant dramatically. Lubrication with higher viscosity usually has a better bearing capacity. At the same time, viscosity is also one of the most critical aspects to influence the friction coefficient. High-viscosity lubricants will generate significant heat and friction; thus, it is important to know the influence of nanomaterial additives on the lubricant viscosity.

When there is relative motion between the friction pairs, the adhesion force of the lubricant to the friction surface and the intermolecular force within the fluid will lead to the shear deformation of the fluid. The viscosity is used to describe the ability of a fluid to resist the shear deformation and the internal friction when flowing. From the description of the viscosity above, one can infer that a fluid's viscosity changes when a fluid is sheared at a different speed. At the same time, the fluid viscosity changes with temperature and pressure. From a molecular perspective, the fluid consists of many molecules in a state of irregular motion, and the viscosity reflects the intermolecular force and molecular momentum. When the temperature increases, the velocity of molecules and the distance between molecules increase, leading to a reduction of the intermolecular force. That is the reason why viscosity changes with temperature. The viscosity index (VI) defines the viscosity of a fluid changed with temperature.

The addition of nanomaterials into the lubricant significantly affects the molecular motion and intermolecular forces of the lubricant. The change in molecular motions and forces will result in the change of rheological properties of the lubricant on the macroscale. Worldwide, researchers have studied the viscosity of lubricants with different nanomaterials. As the investigated tested lubricant additives and the base lubricants are very different, the viscosity data from different researches cannot be compared. However, the laws of viscosity change after the addition of nanomaterial additives are the same. The rheology behavior of silica nanoparticles dispersed

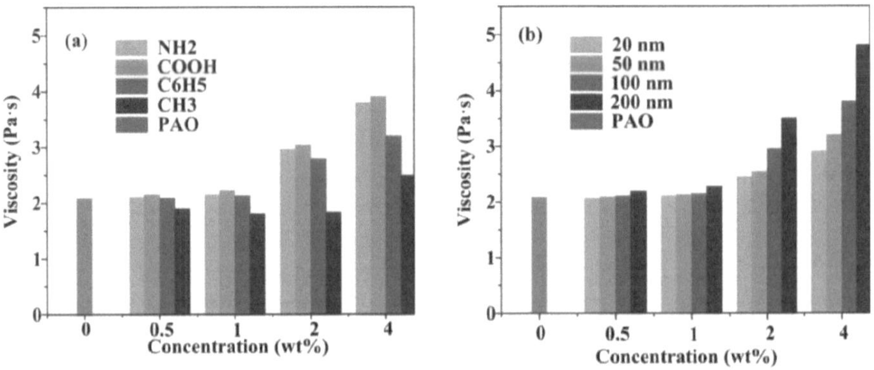


**FIGURE 2.7** The viscosity of PAO 100 with different kinds of silica nanoparticles. (Images were taken from Ref. [12] with the permission of Elsevier.)

in poly-alpha-olefin (PAO) 100 is taken as an example to demonstrate the influence of nanoadditives on the lubricant. In Figure 2.7, when nanoparticles are added to a lubricant, the viscosity of the lubricant increases with the increase of particle concentration and particle size. This phenomenon can be attributed to the influence of nanomaterials on the flow of PAO organic chains. Nanoadditives with a larger diameter and higher concentration would block the flow of PAO organic chains more seriously, leading to an increase of viscosity. The dispersity of the nanomaterial additive would also influence the viscosity of the lubricant. The nanoadditives with poorer dispersion have a more significant influence on the viscosity of the lubricant, and they would form large clusters in the lubricant. The bigger the additive cluster, the harder the lubricant's organic chains flow, and the higher the viscosity. Furthermore, the VI of the lubricant changes slightly after the addition of nanolubricant additives [10,11]. Although the rheology of a lubricant with nanoparticles is very important for its application, studies on viscosity–temperature and viscosity–pressure properties of lubricants with nanoadditives are still at the beginning.

### 2.3.3 The Anti-wear and Friction Reduction Properties of Nano Lubricant Additives

A nanomaterial additive is considered to be an essential direction for the future development of friction modifiers. The anti-wear and friction reduction properties of nanomaterials are the research focuses of worldwide scientists. There are four lubrication regimes—hydrodynamic, elastohydrodynamic, mixed, and boundary. For the hydrodynamic and elastohydrodynamic lubrication, the lubrication film is thick enough to protect the friction pair surface from contacting; thus, friction and wear can be maintained low. In the mixed lubrication and boundary lubrication regime, the lubrication film is not thick enough to support the friction pairs; direct contact happens, leading to high friction and wear. As the gaps between the contacting surface of boundary and mixed lubrication are at the nanoscale, it is believed that the nanomaterials can successfully enter the gap and play a role in reducing friction and wear.

With the development of materials science and characterization techniques, many new materials with different sizes, structures, and functional groups are synthesized and tested with their tribology properties for metallic and non-metallic friction pairs. Nanomaterials are dispersed into mineral oil, synthetic oil, vegetable oil, and water as lubricant additives. Table 2.2 summarizes some of the main studies on the tribological performance of nanomaterials. Almost all the listed studies report the improvement of lubricants in friction reduction and anti-wear performance after the addition of nanomaterials. For the majority of the studies, nanomaterials are added in a low concentration between 0.1% and 10%.

From Table 2.2, several nanomaterials can be used as lubricant additives, and they perform differently. Limited studies are focusing on the difference between different nanoadditives. At the same time, it is challenging to compare the performance of

**TABLE 2.2**
**Tribological Performances of Typical Nanomaterials When Used as Lubricant Additives**

| Category | Nanomaterial | Size | Concentration | Lubricant | Tribological Performance | Reference |
|---|---|---|---|---|---|---|
| Oxides | $SiO_2$ | 100 nm | 1 wt% | PAO 100 | Friction and wear reduced by 40% and 60%, respectively. | [13] |
| | $SiO_2$ | 100 nm | 3~5 wt% | Water | Friction and wear reduced by 38% and 80%, respectively. | [14] |
| | $TiO_2$ | 30 nm | 1 wt% | Water | Extreme pressure performance improved, friction reduced by 36%. | [15] |
| | $TiO_2$ | 20~25 nm | 1 wt% | Mineral oil | Stably dispersed and slightly reduced the friction coefficient. | [16] |
| | $Al_2O_3$ | 80 nm | 0.1 wt% | Mineral oil | Friction and wear reduced by 17.61% and 41.75%, respectively. | [17] |
| | CuO | 50 nm | 2 wt% | PAO 8 | Friction and wear reduced by 18 and 14%, respectively. Load-carrying capacity increased by 273%. | [18] |
| | $Fe_3O_4$ | 10 nm | 0.5, 1, 2 wt% | Liquid paraffin | Friction and wear reduced by 41.8%, 51.5% and 64.7%, respectively. | [19] |

*(Continued)*

**TABLE 2.2 (*Continued*)**
**Tribological Performances of Typical Nanomaterials When Used as Lubricant Additives**

| Category | Nanomaterial | Size | Concentration | Lubricant | Tribological Performance | Reference |
|---|---|---|---|---|---|---|
| Sulfides | $MoS_2$ | 90 nm | 1 wt% | EOT5# engine oil | Friction and wear reduced by 30% and load-carrying capacity improved. | [20] |
| | $MoS_2$ sheet | 20~150 nm | 1 wt% | Liquid paraffin | Load-carrying capacity improved significantly from less than 50 N to more than 2000 N | [21] |
| | IF-$WS_2$ | 120 nm | 1 wt% | Liquid paraffin | Load-carrying capacity improved significantly. | [22] |
| Metal | Cu | 44.7 nm | 0.8 wt% | SF15W/40 oil | Friction and wear reduced by 14.08% and 65.60%, respectively. | [23] |
| | Ag | 6~7 nm | 2 wt% | Multialkylated cyclopentanes | Friction and wear reduced by 17%. | [24] |
| | Fe | 15~50 nm | -- | Mineral oil | Friction and wear reduced by 30% and 55%, respectively. | [25] |
| | Pd | 8 nm | 0.2 wt% | Liquid paraffin | Friction coefficient can be reduced by 30%. | [26] |
| Carbon | Graphene | 1 nm thickness | 110 μg/mL | Water | Friction and wear reduced by 81.3% and 61.8%, respectively. | [27] |
| | Graphene oxide | -- | 1 wt% | Water | Friction coefficient reduced by 57%. | [28] |
| | Carbon nanotubes | 85 nm | 0.01 wt% | Water | Friction coefficient is 0.182, reduced by 11.65%. | [29] |
| | Carbon nanotubes | 85 nm | 0.01 wt% | Oil | Friction coefficient is 0.069, reduced by 15.85%. | [29] |
| | Carbon spheres | 100~500 nm | 3 wt% | SW30 engine oil | Friction and wear reduced by 10%~25% | [30] |

different materials from different studies, because friction conditions, such as friction pair materials, contacting form, base lubricant, sliding speed, and pressure, are very different in each study. However, the influence of nanomaterial parameters for the same kind of nanomaterial on its lubricant performance is widely studied and it is found that they influence the performance significantly. For example, the concentration, size, and surface group play essential roles in the anti-wear and friction reduction properties of nanomaterials.

Silica nanoparticles used as aqueous lubricant additives are selected and presented to show the anti-wear and friction reduction performance of nano lubricant additives. Figure 2.8 shows the tribological performances of silica nanoparticles with different concentrations (0-7 wt%), sizes (20, 50, 100, and 200 nm), functional groups, and under different working conditions (15, 30, and 60 N and 0.25 and 0.5 m/s, respectively) when used as aqueous lubricant additives. The base lubricant used in the experiment is deionized water (DW). The tribological properties are tested by the

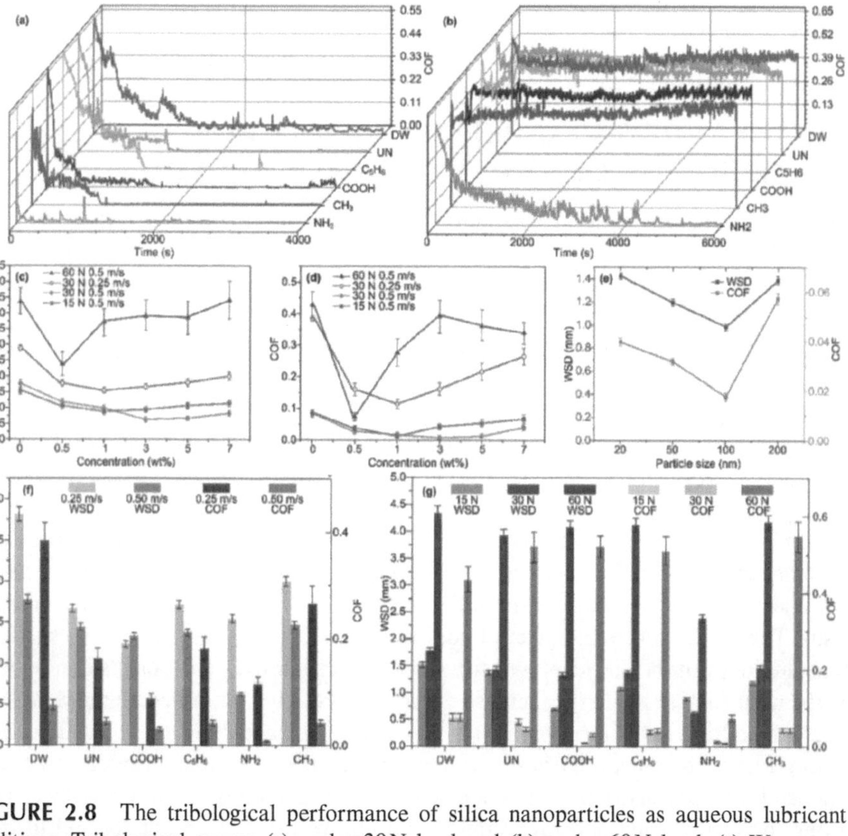


**FIGURE 2.8** The tribological performance of silica nanoparticles as aqueous lubricant additives. Tribological curves (a) under 30 N load and (b) under 60 N load. (c) Wear scar diameter and (d) friction coefficient of $NH_2$ with different concentrations. (e) Tribological properties of $NH_2$ with different particle sizes. Tribological properties of different functional nanoparticles (f) at different sliding speeds and (g) under different loads. (Images were taken from Ref. [9] with the permission of the American Chemical Society.)

ball on the plate test method using a $Si_3N_4$ ball. Unmodified (UN), amino-modified ($NH_2$), carboxyl-modified (COOH), phenyl-modified ($C_6H_5$), and octyl-modified ($CH_3$) silica nanoparticles were tested.

Both friction and wear reduce dramatically after the addition of nanoparticles. The friction coefficient in Figure 2.8a and b highlights that the running-in time is reduced after the addition of silica nanoparticles as lubricant additives. Concentration is a critical parameter for nanoadditives. With the increase of nanoparticle concentration, the lubrication performance improves at first and then worsens. When the nanoadditive is added with a low load, it could not function properly. When too much nanoadditive is added, nanomaterials aggregate easily and harm the wear surface. The lubrication performances of nanoparticles with different surface groups, sizes, and working conditions are very different. For example, the optimal concentration for 60 N load is 0.5 wt%, lower than the optimal concentration of 15 and 30 N (1 wt%). These nanomaterial parameters are essential for lubrication performance. Optimal parameters of the same nanomaterial additives are usually different when applied under different lubrication conditions. Experiments before the application are fundamental to find out the optimal parameters for nanolubricant additives.

## 2.4 THE ANTI-WEAR AND FRICTION REDUCTION MECHANISM OF NANO LUBRICANT ADDITIVES

As demonstrated in Section 1.3, nanomaterials could improve the anti-wear and friction reduction properties of lubricants. This section discusses the friction mechanism of nanomaterials to better understand the lubrication behavior of nanomaterial additives.

The contacting surfaces will wear significantly when sliding with each other under both the boundary and mixed lubrication. Many microscratches will appear on the wear surface; surface bumps and grooves would increase the wear, thus increasing the friction. Advanced characterization methods, such as SEM, energy dispersive spectroscopy, and X-ray photoelectron spectroscopy, allow examining the wear surface systematically after tribology tests to find the clues of the nanoadditive friction mechanism.

Based on the current research results, the four types of friction mechanisms of nanomaterials are rolling effect, protecting effect, mending effect, and polishing effect. Figure 2.9 shows a simplified schematic diagram of the friction mechanism. The addition of nanomaterial additives to a lubricant will repair the surface. Nanoadditives in the lubricant will fill into the microgrooves and defects that appear on the wear surface. With the increase of the nanoadditives deposited on the wear surface, the additives will form protecting films and reduce friction and wear. The protecting film is the most important way for nanoadditives to reduce friction and wear. Nanoadditives can also perform as abrasive particles and polish the wear surface. The filling, protecting, and polishing effect can improve the surface quality and form a smooth surface, which is critical for reducing friction and wear. According to some studies, a nanoadditive can fill the gap between the contact surface and rolling like the function of bearing balls in bearings [31–33]. However, the rolling effect is hard to prove because the wear process takes place in a confined space and is hard to observe.

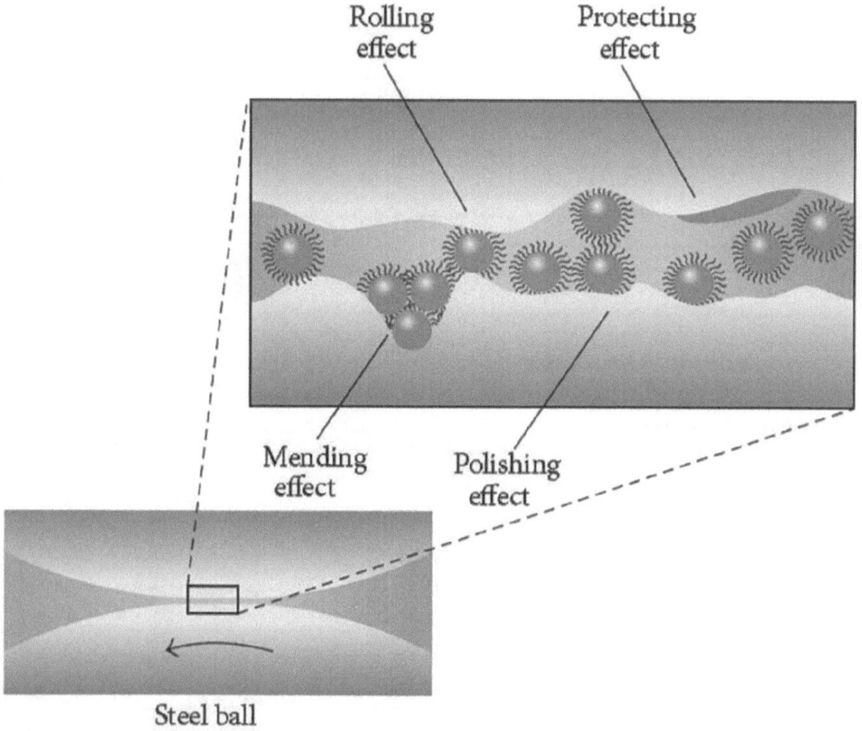


**FIGURE 2.9** The schematic diagram of the nanomaterials friction mechanism. (Images were taken from Ref. [4] with the permission of Hindawi.)

As demonstrated in Section 2.3, the parameters of a nanomaterial influence its tribological performance significantly. In order to better understand the influence mechanism of the nanomaterial parameters on the tribological performance, typical examples were selected for the influence of material, structure, and surface functional group (see Figure 2.10).

From the description of the nanomaterial friction mechanism, nanoadditives play a crucial role in protecting the wear surface for reducing friction and wear. The successful protection of the wear surface, including filling the grooves, forming a protecting film, and polishing the wear surface, is directly related to the surface adsorption of nanoadditives. Surface functional groups are found to have a significant impact on the dispersity of lubricant and adsorption ability on the wear surface. For example, from Figure 2.10a, the chelation effect between metal and amino-ending groups provides the amino-functional groups with better adsorption on the metal surface [34]. SEM and EDS analyses highlight that the nanomaterial surface with high polarity has better adsorption on the wear surface [13]. However, the high polarity surface may lead to severe aggregation of nanomaterials when dispersing into a lubricant, which is bad for the dispersity and stability of nanomaterials as lubricant additives.

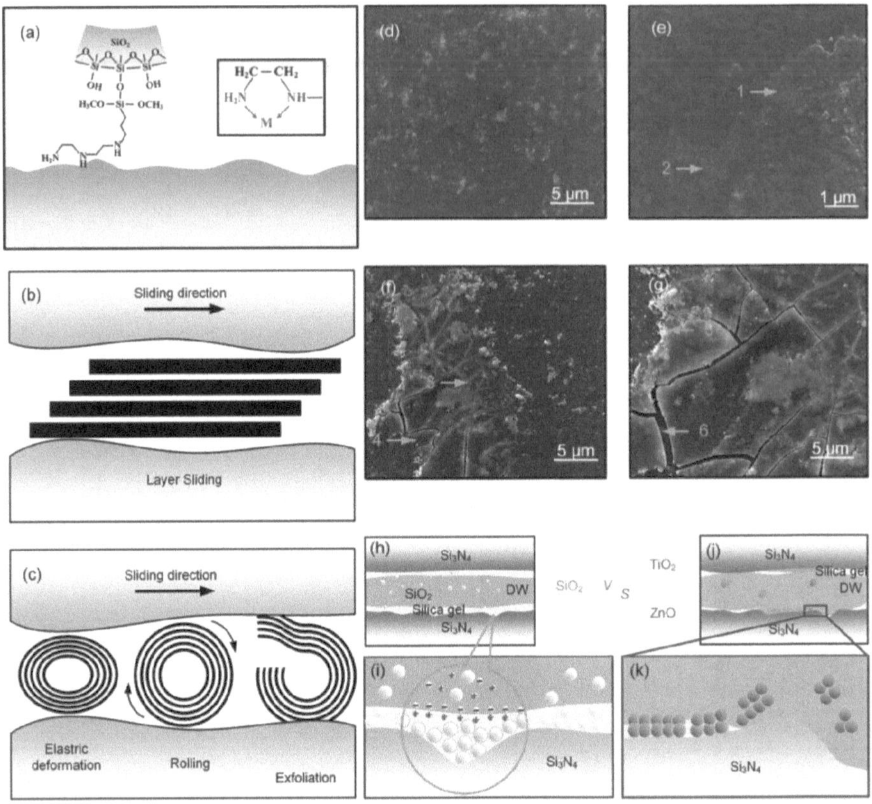


**FIGURE 2.10** The influence of the parameters of nanoadditives on lubrication behaviors: (a) adsorption of surface functional groups on the wear surface; the lubrication mechanism of nanomaterials with (b) nanosheet structures and (c) nanosphere structures; (d) and (e) wear surfaces of ceramics lubricated by $SiO_2$ nanoparticles; (f) and (g) wear surfaces of ceramics lubricated by ZnO nanoparticles; (h) and (i) anti-wear and friction reduction mechanism of $SiO_2$ nanoparticles; (j) and (k) tribological behavior of $TiO_2$ and ZnO nanoparticles. (Images from (d) to (k) are taken from Ref. [35] with permission of Elsevier.)

The structure of nanomaterials may change the friction mechanism, resulting in different tribological performances. Figure 2.10b and c show the friction mechanism of the same material with different nanostructures. Materials with nanosheet and nanosphere morphologies can be synthesized by different processes for materials with a layer structure, such as $MoS_2$ and graphite. For nanosheets, the primary friction mechanism is shearing and sliding of the layer structure material between the contact surfaces. The weak Van der Waals force between layers makes the surface easy to slide, thus reducing friction and wear. However, for nanosphere structures, the friction mechanism changes from sliding to rolling, supporting, and exfoliating. Nanospheres will support the contact surface and deform, roll between the gap like bearings, and exfoliate nanosheets. Those nanosheets exfoliated from a nanosphere will further perform as lubricant additives.

The material type of nanoadditives is also important for the lubrication performance, for example, when using $SiO_2$, ZnO, and $TiO_2$ as aqueous additives for $Si_3N_4$ ceramic lubrication. Figure 2.10d–g shows the wear surfaces after tribology tests. The wear surfaces of ceramics lubricated with $SiO_2$ and ZnO are different. When lubricated with an aqueous lubricant, $Si_3N_4$ ceramics reacts with water and generates silica gel on the wear surface. When adding different kinds of nanoparticles as lubricant additives, $SiO_2$ nanoparticles can form a homogeneous surface film on the wear surface of ceramics, while the surface films formed by ZnO and $TiO_2$ are thick and with cracks. The wear surface of $SiO_2$ is pretty smooth while the wear surface of $TiO_2$ is rough, with defects and grooves. According to the literature, the ceramics nanoparticles have a good tribological performance for ceramics lubrication [9,35]. Furthermore, the metal and metal oxide nanomaterials are reported to form a good surface film when used in metal friction pairs [7,36,37]. The compatibility between the additive material and the friction pair material leads to a totally different lubrication behavior, resulting in different tribological performances. Here we only listed a part of the nanomaterial parameters; the influence of parameters on tribology properties and their mechanisms are still studied by researchers worldwide.

## 2.5 CONCLUSIONS

During the past three decades, researchers worldwide have extensively investigated nano lubricant additives. Oxides, sulfides, metals, and carbon materials are prepared with the size range from 1 to 500 nm and tested with their tribological performance. This study provides examples of the preparation method for the most commonly used nanomaterials. The review of test results from different studies suggested that nanoparticles change the rheological properties and improve the anti-wear and friction reduction properties of lubricants at low concentrations. The parameters of nanomaterial additives, such as particle size, concentration, and surface modification, will influence the tribology performance of lubricants significantly. The friction mechanisms of nano lubricant additives can be summarized as mending, rolling, sliding, polishing, and protecting, while some of the proposed friction mechanisms are hard to verify. With the development of characterization technology, more details and new findings of the friction mechanism will be discovered.

New nanomaterials are continually emerging and applied in the tribology field, such as metal–organic framework nanoparticles and black phosphorus nanosheets. Tribologists are discovering the tribological performance of new nanomaterials and their friction mechanism as lubricant additives. Smart and green nanomaterials will be the future development direction of nano lubricant additives. However, the stability of nanoadditives in the lubricant is still a bottleneck for their commercial application. More work should be done on improving the dispersity and stability of nanomaterials in lubricants in the future.

## REFERENCES

1. Spikes, Hugh. 2015. "Friction Modifier Additives." *Tribology Letters* 60 (1): 1–31. doi:10.1007/s11249-015-0589-z.

2. Stöber, Werner, Arthur Fink, and Ernst Bohn. 1968. "Controlled Growth of Monodisperse Silica Spheres in the Micron Size Range." *Journal of Colloid and Interface Science*. doi:10.1016/0021-9797(68)90272-5.
3. Murphy, Catherine J, Tapan K. Sau, Anand M. Gole, Christopher J. Orendorff, Jinxin. Gao, Linfeng Gou, Simona E. Hunyadi and Tan Li. 2005. "Anisotropic Metal Nanoparticles: Synthesis, Assembly, and Optical Applications." *Journal of Physical Chemistry B* 109: 13857–70. doi:10.1021/jp0516846.
4. Sui, Tianyi, Baoyu Song, Feng Zhang, and Qingxiang Yang. 2015. "Effect of Particle Size and Ligand on the Tribological Properties of Amino Functionalized Hairy Silica Nanoparticles as an Additive to Polyalphaolefin." *Journal of Nanomaterials* 2015: 1–9. doi:10.1155/2015/492401.
5. Shi, Yanbin, Zhaobing Cai, Jibin Pu, Liping Wang, and Qunji Xue. 2019. "Interfacial Molecular Deformation Mechanism for Low Friction of $MoS_2$ Determined Using ReaxFF-MD Simulation." *Ceramics International* 45 (2): 2258–65.
6. Maharaj, Dave, and Bharat Bhushan. 2015. "Nanomechanical Behavior of $MoS_2$ and $WS_2$ Multi-Walled Nanotubes and Carbon Nanohorns." *Scientific Reports* 5: 8539. doi:10.1038/srep08539.
7. Kumara, Chanaka, et al. 2019. "Synergistic Interactions Between Silver and Palladium Nanoparticles in Lubrication." *ACS Applied Nano Materials* 2: 5302–09. doi:10.1021/acsanm.9b01248.
8. Gadipelli, Srinivas, and Zheng Xiao Guo. 2015. "Progress in Materials Science Graphene-Based Materials: Synthesis and Gas Sorption, Storage and Separation." *Journal of Progress in Materials Science* 69, Elsevier Ltd: 1–60. doi:10.1016/j.pmatsci.2014.10.004.
9. Lin, Bin, et al. 2019. "Excellent Water Lubrication Additives for Silicon Nitride To Achieve Superlubricity under Extreme Conditions." *Langmuir* 35: 14861–69. doi:10.1021/acs.langmuir.9b02337.
10. Sepyani, Kamran, et al. 2017. "An Experimental Evaluation of the Effect of ZnO Nanoparticles on the Rheological Behavior of Engine Oil." *Journal of Molecular Liquids* 236, Elsevier B.V.: 198–204. doi:10.1016/j.molliq.2017.04.016.
11. Sanukrishna, Sathyadevan Shyamala, et al. 2018. "Experimental Investigation on Thermal and Rheological Behaviour of PAG Lubricant Modified with $SiO_2$ Nanoparticles." *Journal of Molecular Liquids* 261: 411–22. doi:10.1016/j.molliq.2018.04.066.
12. Sui, Tianyi, Mei Ding, Chunhui Ji, Shuai Yan, Jinhua Wei, Anying Wang, Feifei Zhao, and Jixiong Fei. 2018. "Dispersibility and Rheological Behavior of Functionalized Silica Nanoparticles as Lubricant Additives." *Ceramics International* 44 (15): 18438–43. doi:10.1016/j.ceramint.2018.07.061.
13. Sui, Tianyi, Baoyu Song, Feng Zhang, and Qingxiang Yang. 2016. "Effects of Functional Groups on the Tribological Properties of Hairy Silica Nanoparticles as an Additive to Polyalphaolefin." *RSC Advances* 6 (1): 393–402. doi:10.1039/C5RA22932D.
14. Ding, Mei, Bin Lin, Tianyi Sui, Anying Wang, Shuai Yan, and Qiang Yang. 2018. "The Excellent Anti-Wear and Friction Reduction Properties of Silica Nanoparticles as Ceramic Water Lubrication Additives." *Ceramics International* 44 (12): 14901–6. doi:10.1016/j.ceramint.2018.04.206.
15. Gao, Yongjian, Guoxu Chen, Ya Oli, Zhijun Zhang, and Qunji Xue. 2002. "Study on Tribological Properties of Oleic Acid-Modified $TiO_2$ Nanoparticle in Water." *Wear* 252 (5–6): 454–58. doi:10.1016/S0043-1648(01)00891-2.
16. Ingole, Sudeep, Archana Charanpahari, Amol Kakade, Suresh S. Umare, Dhananjay V. Bhatt, and Jyoti Menghani. 2013. "Tribological Behavior of Nano $TiO_2$ as an Additive in Base Oil." *Wear* 301 (1–2): 776–85. doi:10.1016/j.wear.2013.01.037.
17. Radice, Simona, and Stefano Mischler. 2006. "Effect of Electrochemical and Mechanical Parameters on the Lubrication Behaviour of Al2O3 Nanoparticles in Aqueous Suspensions." *Wear* 261 (9): 1032–41. doi:10.1016/j.wear.2006.03.034.

18. Peña-parás, Laura, Jaime Taha-tijerina, Lorena Garza, Remigiusz Michalczewski, and Carolina Lapray. 2015. "Effect of CuO and $Al_2O_3$ Nanoparticle Additives on the Tribological Behavior of Fully Formulated Oils." *Wear* 333 (2015): 1256–61. doi:10.1016/j.wear.2015.02.038.
19. Zhou, Guanghong, Yufu Zhu, Xiangming Wang, Mujian Xia, Yue Zhang, and Hongyan Ding. 2013. "Sliding Tribological Properties of 0.45% Carbon Steel Lubricated with Fe3O4 Magnetic Nano-Particle Additives in Base Oil." *Wear* 301 (1–2): 753–57. doi:10.1016/j.wear.2013.01.027.
20. Xie, Hongmei, Bin Jiang, Junjie He, Xiangsheng Xia, and Fusheng Pan. 2016. "Lubrication Performance of $MoS_2$ and $SiO_2$ nanoparticles as Lubricant Additives in Magnesium Alloy-Steel Contacts." *Tribology International* 93: 63–70. doi:10.1016/j.triboint.2015.08.009.
21. Chen, Zhe, Xiangwen Liu, Yuhong Liu, Selda Gunsel, and Jianbin Luo. 2015. "Ultrathin $MoS_2$ Nanosheets with Superior Extreme Pressure Property as Boundary Lubricants." *Scientific Reports* 5: 12869. doi:10.1038/srep12869.
22. Rapoport, Lev, Oleg Nepomnyashchy, Igor Lapsker, Armen Verdyan, Alexey Moshkovich, Yishay Feldman, and Reshef Tenne. 2005. "Behavior of Fullerene-like WS2 Nanoparticles under Severe Contact Conditions." *Wear* 259: 703–7. doi:10.1016/j.wear.2005.01.009.
23. Pan, Qiuhong, and Xifeng Zhang. 2010. "Synthesis and Tribological Behavior of Oil-Soluble Cu Nanoparticles as Additive in SF15W/40 Lubricating Oil." *Rare Metal Materials and Engineering* 39 (10): 1711–14. doi:10.1016/S1875-5372(10)60129-4.
24. Ma, Jianqi, Yufei Mo, and Mingwu Bai. 2009. "Effect of Ag Nanoparticles Additive on the Tribological Behavior of Multialkylated Cyclopentanes (MACs)." *Wear* 266: 627–31. doi:10.1016/j.wear.2008.08.006.
25. Padgurskas, Juozas, Igoris Prosyčevas, , Raimundas Rukuiža, Igoris Kreivaitis, and Arturas Kupčinskas. 2012. "Development and Modification of Fe and FeCu Nanoparticles and Tribological Analysis of the Lubricants with Nano-Suspensions," *Industrial Lubrication and Tribology* 64 (5): 253–57. doi:10.1108/00368791211249638.
26. Chen, Shuang, and Weimin Liu. 2006. "Oleic Acid Capped PbS Nanoparticles: Synthesis, Characterization and Tribological Properties." *Materials Chemistry and Physics* 98 (1): 183–89. doi:10.1016/j.matchemphys.2005.09.043.
27. Liang, Shuaishuai, Zhigang Shen, Min Yi, Lei Liu, Xiaojing Zhang, and Shulin Ma. 2016. "In-Situ Exfoliated Graphene for High-Performance Water-Based Lubricants." *Carbon* 96: 1181–90. doi:10.1016/j.carbon.2015.10.077.
28. Elomaa, Oskari, Vivek K. Singh, AjaiIyer, Timo J. Hakala, and Jari Koskinen. 2015. "Graphene Oxide in Water Lubrication on Diamond-like Carbon vs. Stainless Steel High-Load Contacts." *Diamond and Related Materials* 52: 43–48. doi:10.1016/j.diamond.2014.12.003.
29. Antonio, Jesús, Carlos Cornelio, Paula Andrea, Lina Marcela Hoyos-palacio, Javier Lara-romero, and Alejandro Toro. 2015. "Tribological Properties of Carbon Nanotubes as Lubricant Additive in Oil and Water for a Wheel-Rail System." *Integrative Medicine Research* 5 (1): 68–76. doi:10.1016/j.jmrt.2015.10.006.
30. Alazemi, Abdullah A., Vinodkumar Etacheri, Arthur D. Dysart, Lars-Erik Stacke, Vilas G. Pol, and Farshid Sadeghi. 2015. "UltrasmoothSubmicrometer Carbon Spheres as Lubricant Additives for Friction and Wear Reduction." *ACS Applied Materials & Interfaces* 7: 5514–21. doi:10.1021/acsami.5b00099.
31. Lee, Kwangho, et al., 2009. "Understanding the Role of Nanoparticles in Nano-Oil Lubrication." *Tribology Letters* 35 (2009): 127–31. doi:10.1007/s11249-009-9441-7.
32. Si, Lina, et al. June 2012. "Planarization Process of Single Crystalline Silicon Asperity under Abrasive Rolling Effect Studied by Molecular Dynamics Simulation." *Applied Physics A: Materials Science and Processing* 109 (1): 119–26. doi:10.1007/s00339-012-7026-z.

33. Tevet, Ofer, et al. 2011. "Friction Mechanism of Individual Multilayered Nanoparticles." *Proceedings of the National Academy of Sciences* 108 (50): 19901–06. doi:10.1073/pnas.1106553108.
34. Kim, Daniel, and Lynden A. Archer. March 2011. "Nanoscale Organic-Inorganic Hybrid Lubricants." *Langmuir* 27 (6): 3083–94. doi:10.1021/la104937t.
35. Cui, Yuxiao, Mei Ding, Tianyi Sui, Wei Zheng, Guochao Qiao, Shuai Yan, and Xibei Liu. 2020. "Role of Nanoparticle Materials as Water-Based Lubricant Additives for Ceramics." *Tribology International* 142: 105978. doi:10.1016/j.triboint.2019.105978.
36. Padgurskas, Juozas, Raimundas Rukuiza, Igoris Prosyčevas, and Raimondas Kreivaitis. 2013. "Tribological Properties of Lubricant Additives of Fe, Cu and Co Nanoparticles." *Tribology International* 60: 224–32. doi:10.1016/j.triboint.2012.10.024.
37. Hernández Battez, A., Rubén. González, et al. 2008. "CuO, $ZrO_2$ and ZnO Nanoparticles as Antiwear Additive in Oil Lubricants." *Wear* 265 (3–4): 422–28. doi:10.1016/j.wear.2007.11.013.

# 3 Ceramic Matrix High-Temperature Self-Lubricating Materials

*Shengyu Zhu and Dan Wang*
Chinese Academy of Sciences
University of Chinese Academy of Sciences

*Qichun Sun*
Chinese Academy of Sciences

*Jun Cheng and Jun Yang*
Chinese Academy of Sciences
University of Chinese Academy of Sciences

## CONTENTS



DOI: 10.1201/9781003096443-3

## 3.1 INTRODUCTION OF HIGH-TEMPERATURE LUBRICATION

Modern industrial technology urgently needs to solve the high-temperature lubrication problems faced by mechanical systems, such as high-performance aviation engines, hot-end moving parts in aerospace, and heavy metal targets in nuclear energy. However, the friction and wear at high temperatures are very complicated processes, and correspondingly it is extremely difficult to solve the lubrication problems at high temperatures. Since conventional lubricating oil and grease can no longer meet the service requirements due to their oxidation and decomposition above 300°C, solid-lubricating material and technology are an appropriate choice to solve the problem of high-temperature lubrication systems [1–5].

Advanced structural ceramic materials have certain properties, such as high hardness, high strength, low density, and excellent chemical stability, suggesting a wide range of application potential in the field of high-temperature tribology. However, at elevated temperatures, the friction coefficient and wear rate of ceramic materials are relatively high. Under high-temperature environmental conditions, the promising technology for ceramic lubrication should be self-lubricating composite ceramic.

The ceramic materials are mainly composed of ionic bonds and covalent bonds, which make them have the advantages of a high-temperature material matrix. Meanwhile, the intrinsic structure of ceramics leads to severe brittleness and crack susceptibility, thereby causing severe wear during the friction process. Under dry friction conditions, ceramic materials have a relatively high friction coefficient and wear rate, which cannot meet the service requirements, thus limiting their wide application in the field of lubrication. To improve the tribological properties, many efforts are made to design self-lubricating composite ceramics by introducing solid lubricants into the ceramic matrix. The lubrication mechanism of self-lubricating composite ceramics mainly has the following two aspects. (1) One approach is to add a solid lubricant directly into the ceramic matrix. The solid lubricant generates a low-shear lubricating film on the friction surface through the action of friction stress, thereby improving the contact state of the friction interface and realizing the self-lubricating performance of the ceramic material. (2) Another approach is to form the lubricating compound with low shear strength by the tribochemical reaction. Under certain friction conditions, the components in the ceramic matrix composite material can react with the environmental medium to form a substance with lubricating properties and then form a lubricating film on the friction surface to achieve the purpose of lubrication. Based on the above designing approaches, advanced ceramic matrix high-temperature self-lubricating materials are emerging to meet industry application.

Currently, the research on ceramic matrix high-temperature self-lubricating materials has become the forefront topic and cross research in the fields of materials science and tribology. This chapter mainly introduces recent progress in ceramic matrix high-temperature self-lubricating materials, including oxide ceramic matrix, nitride ceramic matrix, carbide ceramic matrix, boride ceramic matrix, and MAX phase ceramic matrix high-temperature self-lubricating materials. Then we summarize the development trends and put forward suggestions for ceramic matrix high-temperature self-lubricating materials.

## 3.2 OXIDE CERAMIC MATRIX HIGH-TEMPERATURE SELF-LUBRICATING MATERIALS

### 3.2.1 $ZrO_2$ Matrix High-Temperature Self-Lubricating Materials

The intrinsic brittleness of ceramic materials is a key factor affecting the safe and reliable service of ceramic matrix composites. Therefore, the strengthening and toughening of ceramics have always been the core issue of ceramic materials research. $ZrO_2$ ceramic is considered an ideal material matrix for high-temperature self-lubricating composites due to the phase transformation toughening effect. The design and preparation of $ZrO_2$ ceramic materials with excellent high-temperature tribological properties have become hotspots in tribological research.

#### 3.2.1.1 Friction Compatibility of Solid Lubricants for $ZrO_2$ Matrix Materials

$ZrO_2$ ceramic has good anti-wear properties at room temperature, but the wear rate of $ZrO_2$ ceramic increases rapidly with increasing temperature, which severely restricts the high-temperature application. To make it reliable at high temperatures, the tribological properties of $ZrO_2$ ceramic materials must be improved. The investigation of the compatibility of solid lubricants with the $ZrO_2$ ceramic is helpful to understand the lubricating mechanism and to provide theoretical guidance for the development of high-temperature self-lubricating materials.

Fluoride, as a kind of solid lubricant with chemical stability, is considered to offer a good high-temperature lubricity, and many studies also prove low friction behavior. It was reported that through using a single $CaF_2$ lubricant for $ZrO_2$ ceramic, the $ZrO_2(Y_2O_3)$–$CaF_2$ composite produces a lubricating $CaF_2$ film that can effectively improve the tribological properties at 600°C [6]. To achieve continuous lubrication at a wide temperature range, it needs to adopt a design of combining a high-temperature solid lubricant and a low-temperature solid lubricant. When studying the combination of $CaF_2$ and graphite, it is found that graphite can effectively improve the friction properties at room temperature, but the oxidation of graphite adversely affects the friction coefficient of the $ZrO_2(Y_2O_3)$–$CaF_2$–graphite composite above 200°C [7].

Although graphite can improve the tribological properties of $ZrO_2$ ceramic at room temperature and 200°C, it is harmful to the tribological properties of $ZrO_2$ ceramic at high temperatures. As for the combination of a noble metal and $CaF_2$, the $ZrO_2$ ($Y_2O_3$)–$CaF_2$–Au composite has a friction coefficient of 0.36–0.50 and a wear rate of $1.6–3.55 \times 10^{-6} mm^3/Nm$. Au is squeezed out of the substrate at low temperatures to form a discontinuous lubricating film, while the plastic deformation of $CaF_2$ and plastic flow of Au play important roles in the formation of the lubricating film above 400°C [8]. The study on the tribological properties of the $ZrO_2(Y_2O_3)$–Mo–$CaF_2$–Ag composite shows that the friction coefficient is very high from room temperature to 600°C, indicating that although soft metal silver is a good lubricant at low temperature, it seems that the insufficient addition of silver does not provide a favorable lubricating effect for the $ZrO_2$ matrix material [9]. If Ag can provide effective lubrication for $ZrO_2$ ceramic, the critical content may approach 35 wt% [10]. When the temperature rises to 800°C and 1000°C, the friction coefficient of the composite quickly decreases to about 0.30, which is mainly due to the generation of the lubricating film containing $CaMoO_4$ that formed on the worn surface by tribo-chemistry of Mo and $CaF_2$ at a high-temperature atmosphere.

$SrSO_4$, as a solid lubricant, can effectively improve the tribological properties of $ZrO_2$ matrix composites at low speed and load [11]. From room temperature to 800°C, the friction coefficient of the $ZrO_2$–$SrSO_4$ matrix composite is between 0.11 and 0.19, and the wear rate is on the order of $10^{-6} mm^3/Nm$. This is attributed to the $SrSO_4$ tribo-layer with low shear strength formed on the worn surface, thereby providing a low friction coefficient at a wide temperature range.

The compatibility of solid lubricants displays that the combination of $CaF_2$ and Ag or Au cannot provide the continuous lubrication for $ZrO_2$ ceramic at a wide temperature range, while the lubricating condition of $SrSO_4$ for $ZrO_2$ ceramic is limited to low load and low speed. Also, this implies that for $ZrO_2$ ceramic, soft metal does not work as an effective solid lubricant, while inorganic salt may be a preferred choice. In addition, based on the advantages of $ZrO_2$ ceramic as a high-temperature matrix, the tribological properties at 800°C and 1000°C should be improved to meet the industrial allowable performance.

#### 3.2.1.2 $ZrO_2$–Mo–CuO High-Temperature Self-Lubricating Composite

The binary oxides MoO–CuO show a promising lubricity at high temperatures due to the low difference of ion potential [12,13]. In the light of the idea that MoO–CuO functions as a high-temperature solid lubricant by the addition of CuO and the oxidative reaction of Mo, a $ZrO_2$–Mo–CuO composite is designed and prepared by powder metallurgy [14].

The results show that at 700°C–1000°C, the friction coefficients of $ZrO_2(Y_2O_3)$–Mo and $ZrO_2(Y_2O_3)$–CuO composites are high to 0.50–0.80, but the friction coefficients of the $ZrO_2$ matrix composites simultaneously containing Mo and CuO are significantly reduced to 0.18–0.30. Meanwhile, the wear resistance of the $ZrO_2(Y_2O_3)$–Mo–CuO composite is remarkably improved by two orders of magnitude compared with the $ZrO_2$ matrix composites added, Mo or CuO. The improvement of high-temperature tribological properties of the $ZrO_2(Y_2O_3)$–Mo–CuO composite is mainly attributed to the formation of a lubricating glaze film consisting of MoO–CuO. The glaze layer is formed by abrasive debris on the worn surface rather than extruded from the substrate.

#### 3.2.1.3 $ZrO_2$–$MoS_2$–$CaF_2$ Self-Lubricating Composite at a Wide Temperature Range

To meet the requirements of engineering application, it is necessary to prepare $ZrO_2$ matrix self-lubricating materials with continuous lubrication properties at a wide temperature range. The combination of $CaF_2$ and $MoS_2$ shows a promising candidate for $ZrO_2$ ceramic at a wide temperature range from room temperature to 1000°C. The $ZrO_2(Y_2O_3)$–$MoS_2$–$CaF_2$ composite prepared by powder metallurgy shows a low friction coefficient and wear rate between room temperature and 1000°C when coupled with $Al_2O_3$ ceramic [15,16].

The $ZrO_2(Y_2O_3)$–$MoS_2$–$CaF_2$ composite shows satisfactory mechanical properties. The microhardness and fracture toughening are HV 824±90 and 6.5±1.4 $MPam^{1/2}$, respectively. The tribological results displayed that $MoS_2$ provides effective lubrication for $ZrO_2$ ceramic from room temperature to 400°C, while at 800°C and 1000°C, the lubricating mechanism is attributed to $CaF_2$ and $CaMoO_4$ formed on worn surfaces due to the tribochemical reaction between $MoS_2$ and $CaF_2$ in the air at elevated temperatures.

When the self-lubricating composite is applied for high-temperature moving parts, the effect of tribo-pair on tribological behavior needs to be studied since friction and wear properties of materials are not their inherent properties. The traditional structural ceramics, such as $Al_2O_3$, $Si_3N_4$, and SiC, are used for high-temperature fields due to their combined performance of high melting point, high-temperature strength, thermal shock resistance, corrosion resistance, oxidation resistance, and structure stability. It is found that the $ZrO_2(Y_2O_3)$–$MoS_2$–$CaF_2$ composite shows different friction and wear properties when coupled with $Al_2O_3$, $Si_3N_4$, and SiC ceramic balls at a wide temperature range.

The friction coefficient of the $ZrO_2(Y_2O_3)$–$MoS_2$–$CaF_2$ composite mated with $Si_3N_4$ and SiC pair is not much different throughout the test temperature range, and both of tribo-pairs have a high friction coefficient of 0.80–0.90 at 600°C. When paired with the $Al_2O_3$ ceramic, the $ZrO_2(Y_2O_3)$–$MoS_2$–$CaF_2$ composite shows a relatively low friction coefficient of 0.2–0.4 from room temperature to 1000°C. When the $ZrO_2(Y_2O_3)$–$MoS_2$–$CaF_2$ composite couples with the $Al_2O_3$ ceramic, from room temperature to 600°C, the wear rate is in the same order of magnitude ($10^{-6}mm^3/Nm$); when paired with $Si_3N_4$ and SiC ceramics, the wear rate of the $ZrO_2(Y_2O_3)$–$MoS_2$–$CaF_2$ composite is increased markedly. Up to 800°C and 1000°C, the wear rate of three tribo-pairs approximate to $10^{-5}mm^3/Nm$. In general, the $ZrO_2(Y_2O_3)$–$MoS_2$–$CaF_2$ composite coupled with the $Al_2O_3$ ceramic shows a favorable wear rate from room temperature to 1000°C. It can be observed that the effect of frictional pairs on lubricating properties is evident at 600°C. The friction coefficient of the $ZrO_2$ matrix composite is lower when coupled with the $Al_2O_3$ ceramic than a Si-containing ceramic. Further analysis of the worn surface indicates that the high friction and wear at 600°C are due to the formation of $SiO_x$ on the worn surface when the $ZrO_2$ matrix composite mates with Si-containing ceramics. $SiO_x$ as a hard abrasive can destroy the lubricating film formed by $CaF_2$, leading to a deterioration of the friction and wear properties.

### 3.2.2 $Al_2O_3$ Matrix High-Temperature Self-Lubricating Materials

Alumina ceramics have great application potential in high-temperature sliding parts due to their advantages such as low density, strong oxidation resistance, and good high-temperature strength. To promote the application of alumina ceramics as high-temperature sliding parts, the design and preparation of alumina matrix high-temperature self-lubricating composites have become an important direction in tribological research in recent years.

The tribological properties of $Al_2O_3$ matrix high-temperature self-lubricating composites have been reported [17,18]. Three $Al_2O_3$ matrix composites ($Al_2O_3$–50 wt% $CaF_2$, $Al_2O_3$–20 wt% Ag-20 wt% $CaF_2$, and $Al_2O_3$–10 wt% Ag-20 wt% $CaF_2$) are prepared by hot-pressing sintering. The experiment (test temperature is 20°C–800°C; pin: $Al_2O_3$ ceramic, disk: $Al_2O_3$ matrix self-lubricating composite) shows that the added solid lubricants of Ag and $CaF_2$ can effectively reduce the friction and wear of the $Al_2O_3$ ceramic between 200°C and 650°C. However, at room temperature and a high temperature of 800°C, the wear rate of the pin and disk is not improved. The formation of a good lubricating film plays an important role in the improvement of friction and wear performance. The formation of the surface lubricating film depends on the plastic deformation of Ag at low temperature, the melt precipitation of Ag at high temperature, and the plastic deformation of $CaF_2$. When the temperature rises to 800°C, the fracture of the lubricating film seriously deteriorates the friction and wear performance. The wear mechanism is transferred from the plastic deformation to brittle fracture. Adding an appropriate amount of a sintering aid can increase its sintering densification and thereby improve its tribological properties. By optimizing the content of the sintering aid and the content of the lubricants, at 650°C, the $Al_2O_3$matrix self-lubricating ceramic exhibits a friction coefficient of 0.3 and a wear rate of $4\times10^{-6}$mm$^3$/Nm. According to the variation of load and speed, the friction and wear diagram can be divided into three interface lubricating areas: no lubricating film, continuous lubricating film, and intermittent lubricating film. At a low load (<10 N), there is no lubricating film formed on the friction interface, which makes the friction and wear very large. At medium load (10–30 N), a continuous lubricating film is formed, mainly composed of molten precipitated Ag and plastically deformed $CaF_2$, the synergistic action of which improves the friction and wear performance. At higher loads (>30 N), the surface film is destroyed. This discontinuous lubricating film does not fully exert its anti-friction effect, leading to the increase in the wear rate and the fluctuation of the friction coefficient.

The effect of sulfates as solid lubricants on tribological properties of $Al_2O_3$ matrix self-lubricating composites has also been studied [19–21]. It is reported that the friction coefficient of $Al_2O_3$–50 wt% $BaSO_4$–20 wt% Ag is relatively low in the temperature range of 200°C–800°C, and above 200°C, there is an Ag lubricating film on the worn surface. Addition of $SrSO_4$ and $PbSO_4$ to the $Al_2O_3$ matrix can result in the same low friction coefficient at 800°C as $Al_2O_3$–50 wt% $BaSO_4$ and $Al_2O_3$–50 wt% $BaCrO_4$. The friction coefficient of the $Al_2O_3$–sulfate composites is 0.42–0.20, and the wear rate is from $10^{-3}$ to $10^{-5}$mm$^3$/Nm.

Bionic-laminated self-lubricating composites are highly attractive materials because of their excellent overall performance and great application prospects [19,-22–28]. It is reported that an $Al_2O_3$/graphite-laminated self-lubricating composite is successfully fabricated by the layer-by-layer method and hot-pressing. The variations of structural parameters (layer number, layer thickness, and layer thickness ratios) cause significant changes in the effective part of the weak interface layer in front of the crack tip and the contribution to the crack propagation energy, thus realizing the optimization of the mechanical properties of materials. The optimal laminated composites show a non-catastrophic fracture characteristic, and the fracture toughness and work of fracture are at least 1.6 and 5.5 times higher than those of monolithic $Al_2O_3$ ceramics, respectively. At room temperature, the friction coefficient and wear rate of the optimal $Al_2O_3$/graphite-laminated composites coupled with an $Al_2O_3$ ball can be reduced to 0.31 and $1.0 \times 10^{-5} mm^3/Nm$, respectively. The $Al_2O_3$/graphite-laminated composites achieve continuous lubrication from room temperature to 800°C by introducing $BaSO_4$ into the weak layers. Furthermore, replacing the graphite phase with $MoS_2$ in the weak layers, the $Al_2O_3/MoS_2$–$BaSO_4$-laminated composite is developed. $BaMoO_4$, formed in situ using a tribochemical reaction, can effectively improve the tribological properties of materials at high temperatures, thus realizing the continuous lubrication within a wide temperature range. The friction coefficient of the optimal laminated materials coupled with an $Al_2O_3$ pin can be controlled in a low range (0.20–0.48), which was 60% lower than that of monolithic alumina ceramics (0.84–1.10) from 25°C to 800°C.

## 3.3 NITRIDE CERAMIC MATRIX HIGH-TEMPERATURE SELF-LUBRICATING MATERIALS

### 3.3.1 Nitride Matrix High-Temperature Self-Adaptive Coatings

Nitride matrix high-temperature self-adaptive coatings, which consist of MeN as a matrix phase and a noble metal as a solid lubricant, exhibit excellent tribological properties from low temperature to high temperature through adjusting surface microstructure and composition, thereby achieving continuous and reliable lubrication at a wide temperature range [29–31]. At present, more attention is paid to the binary coatings, such as TiN–Ag [32,33], CrN–Ag [34], MoN–Ag [35], ZrN–Ag [36], VN–Ag [37], NbN–Ag [38], TaN–Ag [39], and so on. In order to achieve continuous lubrication of nitride coatings from room temperature to 1000°C, Air Force Research Laboratory systematically studied nitride-based temperature-adaptive coatings, such as MoN–Ag, VN–Ag, NbN–Ag, TaN–Ag, and so on. The lubrication temperature range is widened from room temperature to 750°C for MoN–Ag and room temperature to 1000°C for VN–Ag.

#### 3.3.1.1 $Mo_2N$-Ag High-Temperature Self-Adaptive Coating

In the study of $Mo_2N$–$MoS_2$–Ag adaptive coating, it is found that the silver molybdate formed on the frictional surface contributes to the improvement of high-temperature tribological properties [40,41]. The three silver molybdate phases ($Ag_2MoO_4$, $Ag_2Mo_2O_7$, and $Ag_6Mo_{10}O_{33}$) are further explored to analyze the lubricating mechanism. The results show that the three silver molybdate phases have a low friction coefficient of 0.1–0.2, and their melting points are about 500°C–600°C. $Ag_2MoO_4$ has a spinel-type structure, with an Ag layer interposed between the layered structure of $Ag_2O$ and $MoO_3$. During the high-temperature friction process, the weak Ag–O (Ag–O bonding energy 220 kJ/mol, Mo–O bonding energy 560 kJ/mol) bonds are more likely to break, leading to a low friction coefficient. Unlike $Ag_2MoO_4$, $Ag_2Mo_2O_7$ contains the $[Mo_4O_{16}]^{8-}$ chain structure, which is connected by O–Ag–O, thereby obtaining good lubrication. In addition, the silver molybdate phase migrates, agglomerates, and rearranges on the worn surface, resulting in a very low wear rate (the wear of the coating was not detected after 600°C and 300,000 cycles).

#### 3.3.1.2 VN–Ag High-Temperature Self-Adaptive Lubricating Coating

The VN–Ag adaptive high-temperature lubricating coating exhibits an excellent lubricating property at a wide temperature range from room temperature to 1000°C [37, 42]. Combining with lubricating Ag at low to middle temperatures and silver vanadate generated by a tribochemical reaction at middle to high temperatures could be responsible for the excellent tribological performance of the VN/Ag films at a broad temperature range. It is found that besides vanadium oxide, two silver vanadate phases ($AgVO_3$ and $Ag_3VO_4$) are also formed on the worn surface of the VN–Ag adaptive coating. The tests of the thermal stability using in-situ Raman and XRD show that when heated to 450°C, silver vanadate decomposes into Ag and liquid phase ($Ag_3VO_4 \leftrightarrow Ag + liquid$). Due to the low liquid phase shear force, silver vanadate can maintain a low coefficient of friction (0.1–0.25) when paired with $Si_3N_4$ in an atmosphere above 500°C.

#### 3.3.1.3 NbN–Ag and TaN–Ag High-Temperature Self-Adaptive Lubricating Coatings

Referring to the lubrication mechanism of VN–Ag adaptive high-temperature lubricating coating, two self-adaptive NbN–Ag and TaN–Ag coatings are prepared in light of the other two high-melting ternary oxides ($AgNbO_3$ and $AgTaO_3$) [38,39]. When the NbN–Ag adaptive coating is paired with silicon nitride ceramic, the friction coefficient is 0.15–0.3 at the temperature exceeding 700°C in the atmospheric environment. In addition, the tribological properties of the $AgTaO_3$ coating show that $AgTaO_3$ has excellent lubricating properties at high temperatures, with a friction coefficient of about 0.06 at 750°C. After friction at 750°C, $AgTaO_3$ and $Ta_2O_5$ generated on the worn surface can improve the lubrication performance of the coating during high-temperature friction. In order to better understand the lubricating mechanism of $AgTaO_3$ at high temperature, the chemical composition and structure of the worn subsurface of the $AgTaO_3$ coating is analyzed, showing that Ag is wrapped by $Ta_2O_5$ and does not contact $AgTaO_3$ [43,44]. Compared to silver molybdate and silver vanadate, silver tantalate has a stronger Ag–O bonding energy, and accordingly, the breaking energy of Ag–O bonds requires a higher shear stress. However, the $AgTaO_3$ coating also has deficiencies in that the friction coefficient of the $AgTaO_3$ coating depends on the applied load [45]. The experimental results show that at 750°C, when the load increases from 1 N to 10 N, the friction coefficient of the coating coupled with the $Si_3N_4$ ceramic increases from 0.04 to 0.15. Molecular dynamics simulation studies show that under higher loading conditions, Ag is extruded onto the contact surface, covering the $AgTaO_3$ layer. This increases the shear force, resulting in an increase in the friction coefficient. Raman spectroscopy studies confirm that $AgTaO_3$ and $Ta_2O_5$ gradually decrease with the increase in contact pressure.

### 3.3.2 $Si_3N_4$ Matrix High Temperature Self-Lubricating Materials

$Si_3N_4$ ceramics are considered to be one of the most promising structural ceramics due to their high hardness, low density, low thermal expansion, and excellent wide temperature range chemical and mechanical stability. Sialon ceramics, discovered by Oyama in Japan and Jack and Wilson in the United Kingdom from 1971 to 1972 when they studied whether $Al_2O_3$ could be an additive to sinter $Si_3N_4$, has good thermal shock properties and sintering properties [46,47]. Through the efforts of many researchers, Sialon ceramics have been greatly enriched and improved, especially for β-Sialon ceramics and α-Sialon ceramics. In view of their many advantages, silicon nitride-based ceramics ($Si_3N_4$ ceramics and Sialon ceramics) are widely used in sealing parts, engine parts, and high-speed cutting tools. $Si_3N_4$ ceramics face high-temperature friction and wear problems when applied under the above working conditions. Therefore, it is necessary to study the high-temperature tribological problems of silicon nitride ceramics and develop silicon nitride matrix high-temperature self-lubricating materials.

#### 3.3.2.1 $Si_3N_4$–Ag High-Temperature Self-Lubricating Composites

$Si_3N_4$ ceramic undergoes severe friction and wear failure that the friction coefficient and wear rate are 0.9–1.0 and the order of $10^{-3}mm^3/Nm$, respectively, when the temperature reaches 600°C [48]. Therefore, it is very important to study the high-temperature tribological properties of $Si_3N_4$. Although Ag works as an effective solid lubricant, it is difficult to prepare the $Si_3N_4$–Ag composite ceramic by directly adding Ag because of the low melting point of Ag. Recently, a novel sintering technology is developed to fabricate the multiple-phase composite with large differences in melting points among components. Based on the high-temperature solid-phase reaction, $AgNO_3$ is selected as the precursor material of the Ag source, and thereby a $Si_3N_4$–Ag composite ceramic is successfully prepared by spark plasma sintering [48]. $AgNO_3$ is found to decompose at high temperatures and thereby generates Ag in the $Si_3N_4$ matrix. The in-situ formed Ag particles are mainly distributed at the grain boundaries of $Si_3N_4$, and the smallest Ag particles can reach the nanometer level.

The results show that the addition of Ag can significantly improve the tribological properties of the $Si_3N_4$ matrix composite ceramic [49]. As the silver content increases, the friction coefficient decreases significantly, and this downward trend becomes more and more obvious. At 25°C, the friction coefficient of the $Si_3N_4$–Ag composite is 14% lower than that of pure $Si_3N_4$; as the temperature increases to 600°C, the friction coefficient of the $Si_3N_4$–Ag composite is 37% lower than pure $Si_3N_4$. Furthermore, temperature is another factor affecting the tribological performance. It can be seen that the friction coefficients decrease first and then increase with increasing temperatures, and at 200°C, they exhibit the lowest friction coefficient. In addition, with the temperature rising, the mechanical properties of the $Si_3N_4$ matrix ceramic continue to deteriorate, and thereby the wear rate also increases. It is worth noting that the wear rate of the $Si_3N_4$ matrix composite containing Ag is always lower than that of pure $Si_3N_4$ ceramic, which means that the addition of silver can effectively improve the wear resistance of the composite. The favorable wear resistance of the $Si_3N_4$–Ag composite may be attributed to the lubrication and toughening effect of Ag particles.

#### 3.3.2.2 $Si_3N_4$–TiCN High-Temperature Composite

Although the introduction of soft metal Ag with lubricating properties into the $Si_3N_4$ matrix ceramic can improve the high-temperature tribological properties, the hardness of the $Si_3N_4$ matrix is greatly sacrificed. Therefore, it is particularly important to design and prepare $Si_3N_4$-based composites with good comprehensive properties.

$Ti_3SiC_2$ is selected as the precursor material of $TiC_{0.3}N_{0.7}$ to prepare the $Si_3N_4$–$TiC_{0.3}N_{0.7}$ composite [50]. $Si_3N_4$ undergoes a crystalline phase transformation during the sintering process, changing from the original α phase to the final β phase. At the same time, compared with pure $Si_3N_4$, the XRD peaks of $TiC_{0.3}N_{0.7}$ and SiC can be detected in the other samples, and their peak intensities gradually increase with the increase of the $Ti_3SiC_2$ content. This indicates that the decomposition of $Ti_3SiC_2$ happens during the high-temperature sintering process, and the decomposed products further react with $Si_3N_4$ to form $TiC_{0.3}N_{0.7}$. In addition, by magnifying the local XRD peak, another reaction product SiC is detected in the composite ceramic. The possible reaction process can be divided into three stages: the decomposition of $Ti_3SiC_2$, the formation of $TiC_{0.3}N_{0.7}$, and the formation of SiC. First, as the sintering temperature increases, $Ti_3SiC_2$ gradually decomposes, and Si atoms migrate out of the original layered structure, leaving a $TiC_x$ structure with carbon vacancies. As the temperature further increases, $TiC_x$ begins to react with $Si_3N_4$. Specifically, N atoms migrate into the lattice of $TiC_x$ and replace a small amount of C atoms. This process resulted in the formation of $TiC_{0.3}N_{0.7}$. Finally, these replaced C atoms and Si atoms recrystallize, leading to the formation of SiC.

The friction experiments show that as the temperature rises from 25°C to 900°C, the friction coefficient of pure $Si_3N_4$ ceramic increases from 0.44 to 0.99, suggesting that the tribological properties of ceramic deteriorate sharply. However, as for the $Si_3N_4$ matrix composite with the addition of 30 wt% $Ti_3SiC_2$, the friction coefficient changes in the opposite trend, decreasing from 0.57 to 0.43. This result indicates that the introduction of $TiC_{0.3}N_{0.7}$ can play a certain role in reducing friction. With the increase of temperature, the wear rate of $Si_3N_4$ ceramic increases by two orders of magnitude (from $1.17 \times 10^{-5}$ mm³/Nm at 25°C to $1.15 \times 10^{-3}$ mm³/Nm at 900°C). However, with the addition of $TiC_{0.3}N_{0.7}$, this situation has changed significantly. In particular, at 900°C, the wear rate of the $Si_3N_4$–30 wt% $Ti_3SiC_2$ composite is even reduced by three orders of magnitude compared to $Si_3N_4$ ceramic, as low as $6.46 \times 10^{-6}$ mm³/Nm. The enhancement of the tribological properties is mainly due to the improved mechanical properties of the composites and the formation of a tribo-layer with anti-friction and anti-wear effects during the high-temperature friction process.

#### 3.3.2.3 Sialon Matrix Self-Lubricating Composite

β-Sialon ceramic shows an extensive application prospect due to its excellent mechanical properties, good thermal shock resistance and chemical stability, high-temperature resistance, and low density [51–54]. These characteristics make β-Sialon ceramic the most competitive candidate for high-temperature materials and aircraft engines. However, the poor toughness and high-temperature tribological properties of β-Sialon ceramic have further limited its wide applications.

The tribological results of Sialon ceramic at a wide temperature range revealed that β-Sialon ceramic has poor high-temperature tribological properties that the friction coefficient and wear rate are 0.7–0.95 and $10^{-4}$mm$^3$/Nm, respectively [55]. This is mainly due to the fact that β-Sialon ceramic does not have self-lubricating properties at high temperatures. Therefore, it is of great significance to design and prepare β-Sialon ceramic composites with low friction properties at high temperatures.

Graphite can effectively improve the tribological properties of the material at low temperatures. However, graphite is susceptible to oxidation at high temperatures and causes it to lose lubrication; meanwhile, the poor interface between graphite and the matrix severely influences the mechanical properties of the material [56–58]. Based on the previous experimental results, the copper powder can not only improve the high-temperature tribological properties of β-Sialon composite ceramic but also form a Cu–Si interface layer with good interfacial bonding by a high-temperature solid-phase reaction during the sintering preparation process. Therefore, by adjusting the content of adding copper-coated graphite, it is expected to design and prepare a β-Sialon composite with good lubricating properties in a wide temperature range.

The experimental results prove that the addition of copper-coated graphite can effectively improve the tribological properties of β-Sialon composite ceramic. When the copper-coated graphite content is increased to 30 and 40wt%, the composite shows a relatively low friction coefficient from room temperature to 600°C. In particular, the friction coefficient is as low as 0.5 at 600°C, which is 37.5% lower than that of β-Sialon ceramic. When the temperature is increased to 800°C, the friction coefficient of the Sialon–Cu(C) composite is higher than that of β-Sialon ceramic, which is mainly attributed to the severe oxidation of graphite at 800°C. Although the addition of copper-coated graphite causes the increase in wear rate of the β-Sialon matrix composite with the increase of the additive amount at room temperature, the Sialon–Cu(C) matrix composite shows an excellent wear resistance that the wear rate is $10^{-6}$mm$^3$/Nm. Moreover, the high-temperature anti-wear properties of the β-Sialon matrix composite can be effectively improved by adding copper-clad graphite. In particular, from room temperature to 600°C, compared with β-Sialon ceramic, the wear rate of the β-Sialon ceramic composite is reduced by 5–10 times, resulting from the lubricating effect of graphite and the toughening effect of the Cu–Si interface layer.

## 3.4 CARBIDE CERAMIC MATRIX HIGH-TEMPERATURE SELF-LUBRICATING MATERIALS

A silicon carbide ceramic material has a series of excellent properties such as high strength, high hardness, oxidation resistance, wear resistance, corrosion resistance, large thermal conductivity, small thermal expansion coefficient, radiation resistance, and thermal shock resistance. It has a wide range of raw materials and is easy for industrial productions suitable for sliding parts, such as precision bearings, mechanical seals, cutting tools, heat exchangers, and so on. Silicon carbide ceramic has been identified internationally as the fourth-generation basic material after metals, alumina, and cemented carbide. Although silicon carbide ceramic has many excellent properties, it is difficult to meet the requirements of low friction at a wide temperature range. At present, high-temperature tribological research studies on silicon carbide are mainly focused on the friction and wear behavior, and it lacks the self-lubricating behavior under extreme conditions.

Research on the SiC matrix high-temperature self-lubricating material is still in its infancy. The addition of BN, TiC, and $TiB_2$ can improve the tribological properties of silicon carbide ceramics at certain temperatures, but the composite ceramics cannot meet the requirements of self-lubricating materials at a wide temperature range [59,60]. Recently, it is reported that the SiC–Mo–$CaF_2$ solid-lubricating composite prepared by the hot-press sintering method has a low friction coefficient at room temperature and high temperature [61]. By adjusting the content of $CaF_2$ and Mo, the SiC matrix composite has a friction coefficient as low as 0.17 at 1000°C, and the wear rate can be reduced to an order of magnitude of $10^{-5}$mm$^3$/Nm. This is mainly due to the lubricating $CaMoO_4$ produced by the tribochemical reaction between Mo and $CaF_2$ on the worn surface during the high-temperature friction process. Although the addition of metal Mo and high-temperature lubricant $CaF_2$ can provide effective lubrication for SiC ceramic at 800°C and 1000°C, at low temperature, especially at 400°C, the material has a high friction coefficient (up to about 1.0) and severe wear, which makes it impossible to achieve continuous lubrication over a wide temperature range. Therefore, it needs to add a low-temperature solid lubricant to improve the lubrication of SiC ceramic at a wide temperature range. It is found that further addition of $MoS_2$ can effectively improve the tribological properties of the SiC-based composite in the low-temperature range, especially at 400°C; the friction coefficient of the SiC–Mo–$MoS_2$–$CaF_2$ composite is as low as 0.15 and the wear rate approximates to $10^{-6}$mm$^3$/Nm. Moreover, the SiC–Mo–$MoS_2$–$CaF_2$ composite coupled with $Al_2O_3$ ceramic exhibits a friction coefficient of 0.1–0.4 and a wear rate on the order of $10^{-5}$–$10^{-6}$mm$^3$/Nm from room temperature to 1000°C.

## 3.5 BORIDE CERAMIC MATRIX HIGH-TEMPERATURE SELF-LUBRICATING MATERIALS

### 3.5.1 h-BN Matrix High-Temperature Self-Lubricating Materials

h-BN with a laminated crystalline structure, similar to the graphite and $MoS_2$, has been recognized as a potential solid lubricant or material matrix because h-BN layers are bonded by weak van der Waals forces [62–64]. Meanwhile, h-BN exhibits some excellent properties, such as high thermal conductivity, low thermal expansion, excellent chemical stability, and oxidation resistance, which make it a solid-lubricating material suitable for high temperatures [5]. However, only a few works have investigated the tribological properties of h-BN matrix bulk materials. Cao et al. reported the wear behavior of sintered hexagonal boron nitride under atmosphere and water vapor ambience. The results show that the friction coefficient of pure h-BN increases with the temperature increasing from room temperature (0.18) to 400°C (0.58) and then decreases with further increasing the temperature up to 800°C (0.38). In water vapor ambience, the friction coefficients of the h-BN are much lower than those under atmosphere ambience, which are attributed to a lamella-slip of h-BN and the lubricating effect of $H_3BO_3$ formed by the chemical reaction between h-BN and $H_2O$ [65].

As a potential high-temperature solid-lubricating material, it is necessary to investigate the tribological properties of h-BN and its composites. Chen et al. [66] fabricated a h-BN–SiC solid-lubricating composite and investigated the tribological behaviors of the composite from room temperature to 900°C. The experimental results show that the h-BN–SiC solid-lubricating composite has a friction coefficient of 0.2–0.4. Up to 900°C, the friction coefficient is declined to 0.33 and reduced by approximately 38% compared to 0.53 at room temperature. The composite has a maximum friction coefficient of about 0.4 at 400°C. It is also found that increasing the h-BN content is beneficial for the improvement of the friction coefficient. Furthermore, the h-BN-SiC solid-lubricating composite exhibits good lubricating properties in a wide temperature range, which is mainly attributed to weak interlayer bonding and lamella-slip between layers of h-BN. In addition, the results show that the introduction of SiC hard particles as the second phase into the h-BN matrix is helpful to improve the wear resistance of the composites to some extent. However, the h-BN–SiC solid-lubricating composite still exhibits poor wear resistance, and the wear rate reaches the order of magnitude of $10^{-3} mm^3/Nm$.

### 3.5.2 MoAlB Machinable Ceramic

Ceramic materials have great application prospects in the high-temperature tribological fields, due to their high strength, high melting points, and good chemical stability. However, most conventional ceramics, such as $Al_2O_3$, $ZrO_2$, SiC, and $Si_3N_4$, cannot be machined by mechanical or electrical processing methods, which is a fatal limitation to their applications. Therefore, great attention has been paid to the research on machinable ceramics. Recently studies report that the MoAlB ceramic possesses favorable mechanical properties and oxidation resistance [67–71]. Meanwhile, it is worth noting that the MoAlB ceramic can also be easily machined by the wire-cutting electrical discharge machining method. Hence, MoAlB ceramic holds a potential application in high-temperature structural material fields.

Yu et al. studied the high-temperature tribological behaviors of MoAlB ceramic sliding against $Al_2O_3$ and Inconel 718 alloy [72]. The experimental results show that the effects of the test temperature and the counterpart on the tribological behaviors of MoAlB ceramic are evident. The tribological properties of the MoAlB/Inc718 tribo-pair are superior to those of the MoAlB/$Al_2O_3$tribo-pair. The MoAlB/Inc718 tribo-pair demonstrates the relatively low friction coefficient and wear rate. Moreover, for the MoAlB/Inc718 tribo-pair, the friction coefficient is 0.36 at 800°C, and the wear rate is as low as $9.37 \times 10^{-7} mm^3/Nm$ at 600°C. The improvement of friction and wear is mainly attributed to the formation of a tribo-film that has the effects of lubrication and anti-wear during the sliding process at high temperatures.

Benamor et al. also point out that the counterparts have a significant effect on the tribological properties of MoAlB ceramic when sliding against $Al_2O_3$ and 100Cr6 steel [73]. The results show that when MoAlB ceramic is mated with $Al_2O_3$, the friction coefficient increases with increasing load and the wear is highly dependent on the applied load. When MoAlB ceramic is coupled with steel, the friction coefficient decreases with increasing load and the wear rates are low at entire applied loads. Raman spectroscopy indicates that the tribofilms, which are a mixture of $MoO_3$, $B_2O_3$, and $Fe_2O_3$ oxides, are responsible for the improvement of tribological properties of the MoAlB/steel tribo-pair.

### 3.5.3 $AlMgB_{14}$ Superhard Ceramic

$AlMgB_{14}$ had been studied for its framework and photoelectric and thermoelectric properties several decades ago [74–76]. Meanwhile, $AlMgB_{14}$ exhibits superhard (the theoretical value can be as high as 40 GPa), high melting (about 2000°C), lightweight (2.6 g/cm$^3$), and excellent chemical stability [77,78]. Therefore, $AlMgB_{14}$ may be a promising superhard material that can be widely used in the future under some special working conditions.

The tribological behavior of $AlMgB_{14}$ has been investigated under different conditions. It is reported that $AlMgB_{14}$ ceramic shows self-lubricity in an oil or less water environment due to the formation of the lubricating $H_3BO_3$ by tribo-chemistry [79–83], whereas it loses lubricity in large amounts of water because the tribochemical products dissolve in the water completely [84]. Ahmed et al. found that the $AlMgB_{14}$–70 wt% $TiB_2$ composite had high hardness and the highest abrasive resistance. A cutting tool made of $AlMgB_{14}$–70 wt% $TiB_2$ shows low wear due to chipping and little reaction with the Ti-6Al-4V workpiece [85]. Chen et al. prepared a novel $AlMgB_{14}$–Si composite by spark plasma sintering, and its tribological property is investigated under different counterface materials in an ambient environment. The experimental results show that the $AlMgB_{14}$–Si composite exhibits better tribological properties when sliding against 316L steel than when sliding against $Si_3N_4$. The $AlMgB_{14}$–Si composite coupled with 316L steel displays the friction coefficient of 0.19–0.28 and the wear rate of $7.00–8.00 \times 10^{-6}$ mm$^3$/Nm, while when mating with $Si_3N_4$, it presents a relatively high friction coefficient of 0.35–0.46 and a wear rate of $3.80–9.00 \times 10^{-5}$ mm$^3$/Nm. For the former tribo-pair, large numbers of pits formed on the worn surface due to Si grains pull-out, which can store wear debris and is conducive to lower the friction and wear. The latter tribo-pair produces severe fatigue fracture under high Hertz contact stress and thereby aggravates the friction and wear.

The tribological behavior of $AlMgB_{14}$ at room temperature has been performed to understand the lubricating and wear properties, while the high-temperature tribological behavior is seldom investigated. Currently, Chen et al. found that $AlMgB_{14}$ shows excellent wear resistance at room temperature and high temperature (above 900°C), but poor wear resistance at moderate temperature (300°C–700°C) because of the poor fracture toughness. Hence, it is very important to improve the tribological properties of $AlMgB_{14}$ in the middle-temperature stage. $ZrB_2$ is added into the $AlMgB_{14}$ matrix to improve its tribological properties and the friction and wear behaviors of the $AlMgB_{14}$–$ZrB_2$ composite within the temperature range of 25°C–900°C are investigated. $AlMgB_{14}$ shows a high friction coefficient (about 0.8) under temperatures below 700°C (moderate temperature), and then it declines sharply to 0.3 with increasing temperature. However, the friction coefficient of the composite with the addition of $ZrB_2$ is reduced by 20%–50% compared with pure $AlMgB_{14}$ at moderate temperature. The wear rate of pure $AlMgB_{14}$ increases steeply from 25°C to 500°C, reaches the maximum at 500°C ($2.4\times10^{-4}mm^3/Nm$), and then reduces sharply to about $1.4\times10^{-6}mm^3/Nm$ at 700°C until 900°C. As a result, the addition of $ZrB_2$ can obviously reduce the friction coefficient and wear rate of $AlMgB_{14}$. As known, hardness is an important factor to affect the tribological behavior (especially the wear resistance) of a material, while fracture toughness is also a decisive factor in many situations [86,87], especially for brittle ceramics, because they tend to form crack and then produce peeling pits. In this work, though the hardness of the $AlMgB_{14}$–$ZrB_2$ composite declines, the tribological behavior still gets improved to a large degree, which can be attributed to the significant enhancement of the fracture toughness of the composites.

## 3.6 MAX PHASE MATRIX HIGH-TEMPERATURE SELF-LUBRICATING MATERIALS

In recent years, a new type of $M_{n+1}AX_n$ phase ceramics has attracted extensive attention from scholars. $M_{n+1}AX_n$ phase ceramics (abbreviated as MAX phase) are a class of ternary layered compounds, where M represents a class of early transition metals, A stands for a group III or group IV element, and X is a carbon or nitrogen element. According to the different values of n, the $M_{n+1}AX_n$ phase ceramics can be divided into 211, 312, 413, and so on. MAX phase ceramics have excellent combination properties of metals and ceramics. In addition, MAX phase ceramics also have a typical hexagonal layered structure, which is composed of $M_{n+1}X_n$ atomic layers and A atomic layers alternately arranged in this order. This special atomic arrangement makes MAX phase ceramics have the characteristics of a layered structure. Unlike common layered structure lubricants such as graphite and molybdenum disulfide, MAX phase ceramic layered structures do not bind by van der Waals force, but rely on weak metal bonds between M atoms and A atoms, which make MAX phase ceramics have good thermal conductivity and mechanical processing performance.

Because MAX phase ceramics have good comprehensive properties and special layered structures, they are considered as solid-lubricating materials or lubricating additives. As early as 1996, Barsoum et al. reported for the first time that $Ti_3SiC_2$ ceramics may have self-lubricating properties [88]; subsequently, Myhra and Crossley et al. used lateral force microscopy to study the tribological properties of $Ti_3SiC_2$ ceramics and found that their base planes have very low friction coefficient ($\mu\leq5\times10^{-3}$) [89,90]. Based on the above reports, many scholars have conducted a series of studies on the tribological properties of MAX phase ceramics. At present, the most widely studied are $Ti_3SiC_2$ and $Ti_3AlC_2$ ceramics. However, MAX phase ceramics can only exhibit better tribological properties under certain conditions, such as specific frictional pair, high speed, low load, and specific vacuum conditions. In addition, the addition of a second phase can also effectively improve the tribological properties of MAX phase ceramics.

For $Ti_3SiC_2$ ceramic, the lubricating performance of the $Ti_3SiC_2$/diamond frictional pair is significantly better than that of the graphite/diamond frictional pair.

This is due to the self-lubricating film formed during the friction process. Its friction coefficient slightly decreases with increasing load, but the lubrication mechanism is not clear [91]. The microscopic and macroscopic tribological properties of the (0001) crystal epitaxial $Ti_3SiC_2$ thin film prepared by magnetron sputtering are found that under microscopic loading, the friction coefficient is as low as 0.1 when the diamond probe slides on the $Ti_3SiC_2$ thin film; under high load, the friction coefficient obviously increases when $Al_2O_3$ slides against the $Ti_3SiC_2$ thin film, and three-body wear happens. This is because the delaminating cracks formed on the base plane of $Ti_3SiC_2$ cause the surface grain to kink, which leads to an increase in the friction coefficient. When $Ti_3SiC_2$ ceramic is paired with the WC/PbO composite, the tribo-oxide film forms on the friction surface due to the oxidation of the WC/PbO material, which plays an effective lubrication role and reduces the three-body wear at room temperature [92]. However, under certain vacuum conditions, $Ti_3AlC_2$ ceramic exhibits intrinsic self-lubricity independent of the dual material, the friction coefficient is as low as 0.2, and there is almost no wear, but the lubrication mechanism needs to be further studied [93].

Under high-speed sliding conditions (>5 m/s), both $Ti_3SiC_2$ and $Ti_3AlC_2$ ceramics coupled with low-carbon steels exhibit extremely excellent tribological properties. For the frictional pair of $Ti_3SiC_2$/low-carbon steel, a Ti–Si–Fe–O oxide film forms on the frictional surface of $Ti_3AlC_2$ ceramics under high-speed sliding conditions. The friction coefficient and wear rate depend on the morphology of the oxide film formed on the worn surface [94,95]. At a low sliding speed (5 m/s), there is almost no oxide film on the frictional surface of $Ti_3SiC_2$ ceramic; at a high sliding speed (20 and 40 m/s), the formation and coverage of the oxide film vary with the increase in the load, and the tribological characteristics are excellent; when the sliding speed further increases (60 m/s), the oxide film is difficult to form or is easily damaged, leading to the significant increase in friction coefficient. For the $Ti_3AlC_2$/low-carbon steel frictional pair, during the friction process, due to the frictional oxidation reaction, an amorphous friction oxide film composed of Ti, Al, and Fe oxides can be formed on the contact surface and the thickness of the oxide film increases with the increase in sliding speed. During high-speed sliding, the oxide film is in a molten state, so it has a significant self-lubricating effect [96,97].

The high-temperature tribological behavior of MAX phase ceramics shows that at room temperature, MAX phase ceramics exhibit a higher wear rate when paired with nickel-based superalloys, and third-body wear is the cause of the high wear rate; when the temperature is at 550°C, a transferred oxidative film is formed on the contact surface of the friction pair. The transferred film is composed of oxides of Ni-based superalloys, which could play a lubricating role [98,99]. Under high-temperature conditions, when $Ta_2AlC$, $Ti_2AlC$, $Cr_2AlC$, and $Ti_3SiC_2$ ceramics are paired with $Al_2O_3$, an amorphous friction oxide film that is composed of elements M and A oxides can be formed on the worn surface of MAX phase ceramics, which can play an effective anti-wear and anti-friction effect [92]. When $Ti_2SC$ ceramic is paired with $Al_2O_3$, continuous lubrication can be achieved from room temperature to 550°C. This is because an amorphous lubricating oxide film can be formed on the surface of the $Ti_2SC/Al_2O_3$ friction pair during the friction process, which plays a role in continuous lubrication at a wide temperature range [100].

In addition, by combining $Ta_2AlC$ and $Cr_2AlC$ with Ag, high-performance MAX–Ag wear-resistant composites are prepared [101–103]. When the volume fraction of Ag is 20 vol%, $Ta_2AlC$–Ag and $Cr_2AlC$–Ag composites show excellent comprehensive performances. At room temperature, the tensile strength of both composite materials is higher than 150 MPa, and the compressive strength is greater than 1.5 GPa, and the tensile strength is greater than 100 MPa at 550°C. Under the thermal cycling test conditions, the tribological performance of the MAX–Ag/Inc718 frictional pair tends to be stable with the increase of the sliding distance. When sliding against $Al_2O_3$, the wear rate of the MAX–Ag composites increases with increasing temperature. Under high temperature and other thermal cycling conditions, a compacted glaze film formed on the worn surface has a good effect on the substrate's anti-adhesion and lubrication properties. Surface pretreatment may be an effective approach to enhance the tribological performance of the MAX–Ag/nickel-based alloy frictional pair.

## 3.7 CONCLUSIONS AND FUTURE WORKS

Self-lubricating composite ceramics as mechanical moving parts (bearings, bushings, seals, and so on) have a wide range of application prospects in high-end equipment fields, such as aerospace, aviation, marine, and nuclear energy. From the perspective of future development, self-lubricating composite ceramic materials will be used in extreme environments, involving high load, high speed, high temperature, strong corrosion, special media, and atmosphere. However, due to the inherent brittleness of ceramic materials and the decrease in mechanical properties caused by tribological design, their practical applications are largely limited. The reliability and stability of ceramic matrix lubricating materials play important roles in the safe and stable operation of high-end equipment mechanical systems. In the future, ceramic matrix high-temperature self-lubricating materials will develop in the direction of wide temperature-range lubricating materials, multi-function lubricating materials, and intelligent lubricating materials. The following research directions are worthy of attention in the future.

*Design of ceramic matrix lubricating material system*: (1) The materials genome technology can innovate the research and development model of materials by integrating high-throughput computing, high-throughput experiment, and materials database technologies, which greatly accelerates the development of lubricating materials. (2) New material systems, such as layered structural materials and new ceramics, provide more choices for the design of lubricating material systems. Additionally, the combination of a solid lubricant and a tribochemical reaction is still the important strategy to obtain low friction behavior and to achieve lubrication at a wide temperature range.

*Structure/function integrated ceramic matrix lubricating material*: (1) The multi-layer and multi-phase surface design and surface texture technology can build a three-dimensional friction surface, which is helpful to develop a new multifunctional lubricating coating. (2) The biomimetic design concept to the structure and tribological design of ceramic materials can solve the contradiction between the mechanical and tribological properties of conventional ceramic lubricating materials, thereby realizing the structure/lubrication integrated ceramic matrix material.

*Intelligent lubricating material*: The tribo-system can automatically respond to external stimuli (mechanical force, electric field, magnetic field, temperature, and light) and adapt to the working condition of friction performance requirements by the structure adjustment or composition change. Intelligent lubrication broadens the theoretical research system and application scope of high-temperature solid-lubricating materials. The concept of self-adaptive and self-repair ceramic materials provides references for the design idea and feasible technology.

## ACKNOWLEDGMENTS

The authors gratefully acknowledge the financial support from the National Key Research and Development Project of China (No. 2018YFB0703803) and the National Natural Science Foundation of China (51975558 and 51835012).

## REFERENCES

1. Peterson MB, Murray SF, Florek JJ. Consideration of lubricants for temperatures above 1000 F. *Tribol T.* 1959;2:225–34.
2. Sliney HE. Solid lubricant materials for high-temperatures - A review. *Tribol Int.* 1982;15:303–15.
3. Aouadi SM, Luster B, Kohli P, Muratore C, Voevodin AA. Progress in the development of adaptive nitride-based coatings for high temperature tribological applications. *Surf Coat Tech.* 2009;204:962–8.
4. Torres H, Ripoll MR, Prakash B. Tribological behaviour of self-lubricating materials at high temperatures. *Int Mater Rev.* 2018;63:309–40.
5. Zhu S, Cheng J, Qiao Z, Yang J. High temperature solid-lubricating materials: A review. *Tribol Int.* 2019;133:206–23.
6. Ouyang JH, Sasaki S, Umeda K. Low-pressure plasma-sprayed $ZrO_2$-$CaF_2$ composite coating for high temperature tribological applications. *Surf Coat Tech.* 2001;137:21–30.
7. Kong LQ, Zhu SY, Bi QL, Qiao ZH, Yang J, Liu WM. Friction and wear behavior of self-lubricating $ZrO_2(Y_2O_3)$-$CaF_2$-Mo-graphite composite from 20°C to 1000°C. *Ceram Int.* 2014;40:10787–92.
8. Ouyang JH, Sasaki S, Murakami T, Umeda K. The synergistic effects of $CaF_2$ and Au lubricants on tribological properties of spark-plasma-sintered $ZrO_2(Y_2O_3)$ matrix composites. *Mat Sci Eng A-Struct.* 2004;386:234–43.
9. Kong LQ, Zhu SY, Qiao ZH, Yang J, Bi QL, Liu WM. Effect of Mo and Ag on the friction and wear behavior of $ZrO_2(Y_2O_3)$-Ag-$CaF_2$-Mo composites from 20°C to 1000°C. *Tribol Int.* 2014;78:7–13.
10. Ouyang JH, Sasaki S, Murakami T, Umeda K. Tribological properties of spark-plasma-sintered $ZrO_2(Y_2O_3)$–$CaF_2$–Ag composites at elevated temperatures. *Wear.* 2005;258:1444–54.
11. Ouyang JH, Li YF, Wang YM, Zhou Y, Murakami T, Sasaki S. Microstructure and tribological properties of $ZrO_2(Y_2O_3)$ matrix composites doped with different solid lubricants from room temperature to 800°C. *Wear.* 2009;267:1353–60.
12. Erdemir A. A crystal-chemical approach to lubrication by solid oxides. *Tribol Lett.* 2000;8:97–102.
13. Zhu SY, Cheng J, Qiao ZH, Tian Y, Yang J. High temperature lubricating behavior of NiAl matrix composites with addition of CuO. *J Tribol-T ASME.* 2016;138: 031607-1–9.

14. Kong LQ, Bi QL, Zhu SY, Qiao ZH, Yang J, Liu WM. Effect of CuO on self-lubricating properties of $ZrO_2(Y_2O_3)$-Mo composites at high temperatures. *J Eur Ceram Soc.* 2014;34:1289–96.
15. Kong LQ, Bi QL, Niu MY, Zhu SY, Yang J, Liu WM. High-temperature tribological behavior of $ZrO_2$-$MoS_2$-$CaF_2$ self-lubricating composites. *J Eur Ceram Soc.* 2013;33:51–9.
16. Kong LQ, Bi QL, Niu MY, Zhu SY, Yang J, Liu WM. $ZrO_2(Y_2O_3)$-$MoS_2$-$CaF_2$ self-lubricating composite coupled with different ceramics from 20°C to 1000°C. *Tribol Int.* 2013;64:53–62.
17. Jin Y, Kato K, Umehara N. Effects of sintering aids and solid lubricants on tribological behaviours of CMC/$Al_2O_3$ pair at 650°C. *Tribol Lett.* 1999;6:15–21.
18. Jin Y, Kato K, Umehara N. Further investigation on the tribological behavior of $Al_2O_{3\_}$ 20Ag20$CaF_2$ composite at 650°C. *Tribol Lett.* 1999;6:225–32.
19. Song JJ, Hu LT, Qin BF, Fan HZ, Zhang YS. Fabrication and tribological behavior of $Al_2O_3$/$MoS_2$-$BaSO_4$ laminated composites doped with in situ formed $BaMoO_4$. *Tribol Int.* 2018;118:329–36.
20. Murakami T, Ouyang JH, Umeda K, Sasaki S. High-temperature friction properties of $BaSO_4$ and $SrSO_4$ powder films formed on $Al_2O_3$ and stainless steel substrates. *Mater Sci Eng: A.* 2006;432:52–8.
21. Murakami T, Ouyang JH, Sasaki S, Umeda K, Yoneyama Y. High-temperature tribological properties of spark-plasma-sintered $Al_2O_3$ composites containing barite-type structure sulfates. *Tribol Int.* 2007;40:246–53.
22. Fang Y, Fan H, Song J, Zhang Y, Hu L. Surface engineering design of $Al_2O_3$/Mo self-lubricating structural ceramics - Part II: Continuous lubrication effects of a three-dimensional lubricating layer at temperatures from 25 to 800°C. *Wear.* 2016;360–361:97–103.
23. Song JJ, Su YF, Fan HZ, Zhang YS, Hu LT. A novel design to produce high-strength and high-toughness alumina self-lubricated composites with enhanced thermal-shock resistance-Part I: Mechanical properties and thermal shock behavior of $Al_2O_3$/Mo-$Al_2O_3$ laminated composites. *J Eur Ceram Soc.* 2017;37:213–21.
24. Fang Y, Fan HZ, Song JJ, Zhang YS, Hu LT. Surface engineering design of $Al_2O_3$/Mo self-lubricating structural ceramics - Part II: Continuous lubrication effects of a three-dimensional lubricating layer at temperatures from 25 to 800°C. *Wear.* 2016;360:97–103.
25. Fan HZ, Hu TC, Wan HQ, Zhang YS, Song JJ, Hu LT. Surface composition-lubrication design of $Al_2O_3$/Ni laminated composites - Part II: Tribological behavior of $LaF_3$-doped $MoS_2$ composite coating in a water environment. *Tribol Int.* 2016;96:258–68.
26. Song JJ, Zhang YS, Fan HZ, Hu TC, Hu LT, Qu JM. Design of interfaces for optimal mechanical properties in $Al_2O_3$/Mo laminated composites. *J Eur Ceram Soc.* 2015;35:1123–7.
27. Song JJ, Zhang YS, Su YF, Fang Y, Hu LT. Influence of structural parameters and compositions on the tribological properties of alumina/graphite laminated composites. *Wear.* 2015;338:351–61.
28. Fang Y, Zhang YS, Song JJ, Fan HZ, Hu LT. Influence of structural parameters on the tribological properties of $Al_2O_3$/Mo laminated nanocomposites. *Wear.* 2014;320:152–60.
29. Voevodin AA, Fitz TA, Hu JJ, Zabinski JS. Nanocomposite tribological coatings with "chameleon" surface adaptation. *J Vac Sci Technol A.* 2002;20:1434–44.
30. Franz R, Mitterer C. Vanadium containing self-adaptive low-friction hard coatings for high-temperature applications: A review. *Surf Coat Tech.* 2013;228:1–13.
31. Voevodin AA, Muratore C, Aouadi SM. Hard coatings with high temperature adaptive lubrication and contact thermal management: Review. *Surf Coat Tech.* 2014;257:247–65.
32. Muratore C, Voevodin AA, Hu JJ, Zabinski JS. Multilayered YSZ-Ag-Mo/TiN adaptive tribological nanocomposite coatings. *Tribol Lett.* 2006;24:201–6.

33. Köstenbauer H, Fontalvo GA, Mitterer C, Keckes J. Tribological properties of TiN/Ag nanocomposite coatings. *Tribol Lett.* 2008;30:53–60.
34. Kutschej K, Mitterer C, Mulligan CP, Gall D. High-temperature tribological behavior of CrN-Ag self-lubricating coatings. *Adv Eng Mater.* 2006;8:1125–9.
35. Gulbinski W, Suszko T. Thin films of $Mo_2N$/Ag nanocomposite - the structure, mechanical and tribological properties. *Surf Coat Tech.* 2006;201:1469–76.
36. Ju HB, Yu D, Yu LH, Ding N, Xu JH, Zhang XD, et al. The influence of Ag contents on the microstructure, mechanical and tribological properties of ZrN-Ag films. *Vacuum.* 2018;148:54–61.
37. Aouadi SM, Singh DP, Stone DS, Polychronopoulou K, Nahif F, Rebholz C, et al. Adaptive VN/Ag nanocomposite coatings with lubricious behavior from 25 to 1000°C. *Acta Mater.* 2010;58:5326–31.
38. Stone DS, Migas J, Martini A, Smith T, Muratore C, Voevodin AA, et al. Adaptive NbN/Ag coatings for high temperature tribological applications. *Surf Coat Tech.* 2012;206:4316–21.
39. Stone DS, Harbin S, Mohseni H, Mogonye JE, Scharf TW, Muratore C, et al. Lubricious silver tantalate films for extreme temperature applications. *Surf Coat Tech.* 2013;217:140–6.
40. Aouadi SM, Paudel Y, Simonson WJ, Ge Q, Kohli P, Muratore C, et al. Tribological investigation of adaptive $Mo_2N/MoS_2$/Ag coatings with high sulfur content. *Surf Coat Tech.* 2009;203:1304–9.
41. Aouadi SM, Paudel Y, Luster B, Stadler S, Kohli P, Muratore C, et al. Adaptive $Mo_2N/MoS_2$/Ag tribological nanocomposite coatings for aerospace applications. *Tribol Lett.* 2008;29:95–103.
42. Luster B, Stone D, Singh DP, to Baben M, Schneider JM, Polychronopoulou K, et al. Textured VN coatings with $Ag_3VO_4$ solid lubricant reservoirs. *Surf Coat Tech.* 2011;206:1932–5.
43. Gao H, Otero-De-La-Roza A, Gu J, Stone D, Aouadi SM, Johnson ER, et al. (Ag, Cu)-Ta-O ternaries as high-temperature solid-lubricant coatings. *ACS Appl Mater Inter.* 2015;7:15422–9.
44. Stone DS, Gao H, Chantharangsi C, Paksunchai C, Bischof M, Martini A, et al. Reconstruction mechanisms of tantalum oxide coatings with low concentrations of silver for high temperature tribological applications. *Appl Phys Lett.* 2014;105.
45. Stone DS, Gao H, Chantharangsi C, Paksunchai C, Bischof M, Jaeger D, et al. Load-dependent high temperature tribological properties of silver tantalate coatings. *Surf Coat Tech.* 2014;244:37–44.
46. Jack KH, Wilson WI. Ceramics based on Si-Al-O-N and related systems. *Nat-Phys Sci.* 1972;238:28–29.
47. Yoichi O. Solid solution in the ternary system, $Si_3N_4$-AlN-$Al_2O_3$. *Jpn J Appl Phys.* 1972;11:760.
48. Liu JJ, Yang J, Yu Y, Sun QC, Qiao ZH, Liu WM. Self-Lubricating $Si_3N_4$-based composites toughened by in situ formation of silver. *Ceram Int.* 2018;44:14327–34.
49. Liu JJ, Wang ZX, Yin B, Yang J, Sun QC, Liu YL, et al. A novel method to prepare self-lubricity of $Si_3N_4$/Ag composite: Microstructure, mechanical and tribological properties. *J Am Ceram Soc.* 2018;101:3745–8.
50. Liu JJ, Yang J, Zhu SY, Cheng J, Yu Y, Qiao ZH, et al. Temperature-driven wear behavior of $Si_3N_4$-based ceramic reinforced by in situ formed $TiC_{0.3}N_{0.7}$ particles. *J Am Ceram Soc.* 2019;102:4333–43.
51. Yang ZF, Shang QL, Shen XY, Zhang LQ, Gao JS, Wang H. Effect of composition on phase assemblage, microstructure, mechanical and optical properties of Mg-doped sialon. *J Eur Ceram Soc.* 2017;37:91–8.

52. Al Malki MM, Khan RMA, Hakeem AS, Hampshire S, Laoui T. Effect of Al metal precursor on the phase formation and mechanical properties of fine-grained SiAlON ceramics prepared by spark plasma sintering. *J Eur Ceram Soc*. 2017;37:1975–83.
53. Li ZM, Wang ZJ, Zhu MG, Li JF, Zhang ZT. Oxidation behavior of beta-SiAlON powders fabricated by combustion synthesis. *Ceram Int*. 2016;42:7290–9.
54. Huang ZH, Yang JZ, Liu YG, Fang MH, Huang JT, Sun HR, et al. Novel sialon-based ceramics toughened by Ferro-molybdenum alloy. *J Am Ceram Soc*. 2012;95:859–61.
55. Sun Q, Yang J, Yin B, Tan H, Liu Y, Liu J, et al. Dry sliding wear behavior of β-Sialon ceramics at wide range temperature from 25 to 800°C. *J Eur Ceram Soc*. 2017;37:4505–13.
56. Tan H, Wang S, Yu Y, Cheng J, Zhu S, Qiao Z, et al. Friction and wear properties of Al-20Si-5Fe-2Ni-Graphite solid-lubricating composite at elevated temperatures. *Tribol Int*. 2018;122:228–35.
57. Tan H, Wang S, Cheng J, Zhu SY, Yu Y, Qiao ZH, et al. Tribological properties of Al-20Si-5Fe-2Ni-Graphite solid-lubricating composites. *Tribol Int*. 2018;121:214–22.
58. Cui GJ, Niu MY, Zhu SY, Yang J, Bi QL. Dry-sliding tribological properties of bronze-graphite composites. *Tribol Lett*. 2012;48:111–22.
59. Skopp A, Woydt M. Ceramic-ceramic composite-materials with improved friction and wear properties. *Tribol Int*. 1992;25:61–70.
60. Chen ZS, Li HJ, Fu QG, Qiang XF. Tribological behaviors of SiC/h-BN composite coating at elevated temperatures. *Tribol Int*. 2012;56:58–65.
61. Li F, Zhu SY, Cheng J, Qiao ZH, Yang J. Tribological properties of Mo and $CaF_2$ added SiC matrix composites at elevated temperatures. *Tribol Int*. 2017;111:46–51.
62. Zhao Y, Feng K, Yao C, Nie P, Huang J, Li Z. Microstructure and tribological properties of laser cladded self-lubricating nickel-base composite coatings containing nano-Cu and h-BN solid lubricants. *Surf Coat Technol*. 2019;359:485–94.
63. Podgornik B, Kosec T, Kocijan A, Donik Č. Tribological behaviour and lubrication performance of hexagonal boron nitride (h-BN) as a replacement for graphite in aluminium forming. *Tribol Int*. 2015;81:267–75.
64. Kimura Y, Wakabayashi T, Okada K, Wada T, Nishikawa H. Boron nitride as a lubricant additive. *Wear*. 1999;232:199–206.
65. Cao Y, Du L, Huang C, Liu W, Zhang W. Wear behavior of sintered hexagonal boron nitride under atmosphere and water vapor ambiences. *Appl Surf Sci*. 2011;257:10195–200.
66. Chen J, Chen J, Wang S, Sun Q, Cheng J, Yu Y, et al. Tribological properties of h-BN matrix solid-lubricating composites under elevated temperatures. *Tribol Int*. 2020;148:106333.
67. Lu XG, Li S, Zhang WW, Yu WB, Zhou Y. Thermal shock behavior of a nanolaminated ternary boride: MoAlB. *Ceram Int*. 2019;45:9386–9.
68. Bei G, van der Zwaag S, Kota S, Barsoum MW, Sloof WG. Ultra-high temperature ablation behavior of MoAlB ceramics under an oxyacetylene flame. *J Eur Ceram Soc*. 2019;39:2010–7.
69. Ali MA, Hadi MA, Hossain MM, Naqib SH, Islam AKMA. Theoretical investigation of structural, elastic, and electronic properties of ternary boride MoAlB. *Phys Status Solidi B*. 2017;254:1–10.
70. Kota S, Agne M, Zapata-Solvas E, Dezellus O, Lopez D, Gardiola B, et al. Elastic properties, thermal stability, and thermodynamic parameters of MoAlB. *Phys Rev B*. 2017;95:1–11.
71. Kota S, Zapata-Solvas E, Chen YX, Radovic M, Lee WE, Barsoum MW. Isothermal and cyclic oxidation of MoAlB in air from 1100°C to 1400°C. *J Electrochem Soc*. 2017;164:C930–C8.
72. Yu ZG, Tan H, Wang S, Cheng J, Sun QC, Yang J, et al. High-temperature tribological behaviors of MoAlB ceramics sliding against $Al_2O_3$ and Inconel 718 alloy. *Ceram Int*. 2020;46:14713–20.

73. Benamor A, Kota S, Chiker N, Haddad A, Hadji Y, Natu V, et al. Friction and wear properties of MoAlB against $Al_2O_3$ and 100Cr6 steel counterparts. *J Eur Ceram Soc.* 2019;39:868–77.
74. Werheit H, Kuhlmann U, Krach G, Higashi I, Lundstrom T, Yu Y. Optical and electronic-properties of the orthorhombic $MgAlB_{14}$-type borides. *J Alloy Compd.* 1993; 202:269–81.
75. Higashi I, Ito T. Refinement of the structure of $MgAlB_{14}$. *J Less-Common Met.* 1983;92:239–46.
76. Matkovic VI, Economy J. Structure of $MgAlB_{14}$ and a brief critique of structural relationships in higher borides. *Acta Crystall B-Stru.* 1970;B 26:616–621.
77. Cook BA, Harringa JL, Lewis TL, Russell AM. A new class of ultra-hard materials based on $AlMgB_{14}$. *Scripta Mater.* 2000;42:597–602.
78. Zhou YC, Dai FZ, Xiang HM, Feng ZH. Near-isotropic elastic properties and high shear deformation resistance: Making low symmetry and open structured $YbAlB_{14}$, $LuAlB_{14}$ and $ScMgB_{14}$ superhard. *Acta Mater.* 2017;135:44–53.
79. Higdon C, Cook B, Harringa J, Russell A, Goldsmith J, Qu J, et al. Friction and wear mechanisms in $AlMgB_{14}$-$TiB_2$ nanocoatings. *Wear.* 2011;271:2111–5.
80. Lu X, Yao K, Ouyang J, Tian Y. Tribological characteristics and tribo-chemical mechanisms of Al-Mg-Ti-B coatings under water-glycol lubrication. *Wear.* 2015;326:68–73.
81. Cook BA, Harringa JL, Anderegg J, Russell AM, Qu J, Blau PJ, et al. Analysis of wear mechanisms in low-friction $AlMgB_{14}$-$TiB_2$ coatings. *Surf Coat Tech.* 2010;205:2296–301.
82. Chen J, Cheng J, Li F, Zhu SY, Qiao ZH, Yang J. The effect of compositional tailoring and sintering temperature on the mechanical and tribological properties of Cu/$AlMgB_{14}$ composite. *Tribol Int.* 2016;96:155–62.
83. Cheng J, Ma JQ, Li F, Qiao ZH, Yang J, Liu WM. Dry-sliding tribological properties of Cu/$AlMgB_{14}$ composites. *Tribol Lett.* 2014;55:35–44.
84. Chen J, Cheng J, Wang S, Zhu SY, Qiao ZH, Yang J. Self-lubricity and wear behaviors of bulk polycrystalline $AlMgB_{14}$ depending on the counterparts in deionized water. *Tribol Int.* 2018;128:9–20.
85. Ahmed A, Bahadur S, Russell AM, Cook BA. Belt abrasion resistance and cutting tool studies on new ultra-hard boride materials. *Tribol Int.* 2009;42:706–13.
86. Fischer TE, Anderson MP, Jahanmir S. Influence of fracture toughness on the wear resistance of yttria-doped zirconium oxide. *J Am Ceram Soc.* 1989;72:252–7.
87. Sun Q, Yang J, Yin B, Cheng J, Zhu S, Wang S, et al. High toughness integrated with self-lubricity of Cu-doped Sialon ceramics at elevated temperature. *J Eur Ceram Soc.* 2018;38:2708–15.
88. Barsoum MW, ElRaghy T. Synthesis and characterization of a remarkable ceramic: $Ti_3SiC_2$. *J Am Ceram Soc.* 1996;79:1953–6.
89. Myhra S, Summers JWB, Kisi EH. $Ti_3SiC_2$- A layered ceramic exhibiting ultra-low friction. *Mater Lett.* 1999;39:6–11.
90. Crossley A, Kisi EH, Summers JWB, Myhra S. Ultra-low friction for a layered carbide-derived ceramic, $Ti_3SiC_2$, investigated by lateral force microscopy (LFM). *J Phys D Appl Phys.* 1999;32:632–8.
91. Zhang Y, Ding GP, Zhou YC, Cai BC. $Ti_3SiC_2$- a self-lubricating ceramic. *Mater Lett.* 2002;55:285–9.
92. Gupta S, Filimonov D, Palanisamy T, Barsoum MW. Tribological behavior of select MAX phases against $Al_2O_3$ at elevated temperatures. *Wear.* 2008;265:560–5.
93. Ma JQ, Hao JY, Fu LC, Qiao ZH, Yang J, Liu WM, et al. Intrinsic self-lubricity of layered $Ti_3AlC_2$ under certain vacuum environment. *Wear.* 2013;297:824–8.
94. Zhai HX, Huang ZY, Zhou Y, Zhang ZL, Wang YF, Ai MX. Oxidation layer in sliding friction surface of high-purity $Ti_3SiC_2$. *J Mater Sci.* 2004;39:6635–7.

95. Huang Z, Zhai H, Guan M, Liu X, Ai M, Zhou Y. Oxide-film-dependent tribological behaviors of $Ti_3SiC_2$. *Wear.* 2007;262:1079–85.
96. Zhai HX, Huang ZY, Ai MX, Yang Z, Zhang ZL, Li SB. Tribophysical properties of polycrystalline bulk $Ti_3AlC_2$. *J Am Ceram Soc.* 2005;88:3270–4.
97. Huang ZY, Zhai H, Zhou WX, Liu X, Ai MX. Tribological behaviors and mechanisms of $Ti_3AlC_2$. *Tribol Lett.* 2007;27:129–35.
98. Gupta S, Filimonov D, Zaitsev V, Palanisamy T, Barsoum MW. Ambient and 550°C tribological behavior of select MAX phases against Ni-based superalloys. Wear. 2008;264:270–8.
99. Li H, Ren S, Shang J, Lv J. Tribological properties of $Ti_3SiC_2$-Inconel 718 couple at elevated temperatures. *Tribol Chin.* 2013;33:129–34.
100. Gupta S, Amini S, Filimonov D, Palanisamy T, El-Raghy T, Barsoum MW. Tribological behavior of $Ti_2SC$ at ambient and elevated temperatures. *J Am Ceram Soc.* 2007;90:3566–71.
101. Gupta S, Filimonov D, Zaitsev V, Palanisamy T, El-Raghy T, Barsoum MW. Study of tribofilms formed during dry sliding of $Ta_2AlC$/Ag or $Cr_2AlC$/Ag composites against Ni-based superalloys and $Al_2O_3$. *Wear.* 2009;267:1490–500.
102. Filimonov D, Gupta S, Palanisamy T, Barsoum MW. Effect of applied load and surface roughness on the tribological properties of Ni-based superalloys versus $Ta_2AlC$/Ag or $Cr_2AlC$/Ag composites. *Tribol Lett.* 2009;33:9–20.
103. Gupta S, Filimonov D, Palanisamy T, El-Raghy T, Barsoum MW. $Ta_2AlC$ and $Cr_2AlCAg$-based composites - New solid lubricant materials for use over a wide temperature range against Ni-based superalloys and alumina. *Wear.* 2007;262:1479–89.

# 4 Exploration of Bio-Greases for Tribological Applications

*Sooraj Singh Rawat and A. P. Harsha*
Indian Institute of Technology (Banaras Hindu University)

## CONTENTS



## 4.1 INTRODUCTION

Tribopairs have high sliding friction when they interact under unlubricated conditions. Therefore, high frictional heat is generated, which often leads to an increase in wear rate, and a significant amount of energy is required to overcome frictional

DOI: 10.1201/9781003096443-4

losses. Holmberg et al. [1] reported that, on average, a single unit of trucks consumes 1500 liter of diesel per year worldwide to overcome friction. Similarly, it has been reported that 15%–25% of energy is consumed in paper mills to overcome friction [2]. This indicates that a massive amount of energy is dissipated due to frictional losses. Therefore, the application of lubricants/greases is necessary to overcome the friction between the friction pairs. The primary purpose of a lubricant is to control the friction and wear at the interacting surfaces of machinery. A large variety of lubricants available in the form of liquids, semi-solids, solids, or sometimes gases are used to lubricate surfaces. A proper lubrication practice is an essential aspect of any machinery. In a lubricated condition, the lubricant reduces the asperity interactions and reduces the adhesive forces between friction pairs. Adequate lubrication reduces the overall friction in the system as well as minimizes heat generation. A reduction in frictional losses not only decreases the maintenance cost but also reduces the economic losses. The 10% reduction in rolling resistance in a passenger car can reduce 2% in demand for energy [3].

'Grease' is derived from a Latin word 'Crassus' means fat.. The obligation of grease is not unlike lubricant. Grease seems like a 'gel', semi-solid, or a solid product formulated via a homogeneous dispersion of a thickener in a base lube oil. Some additives and fillers are additional materials to strengthen the basic characteristics of grease. Greases are more resistant to water and impersonate seals and protect the surfaces against contamination and dirt compared to lubricating oil. Greases also diminish the splash or leakage tendency of lubricating oil. The viscosity of lube oil is very sensitive to temperature, while the viscosity of grease is less affected by temperature. The grease does not flow itself because of its viscoelastic behavior under the effect of force, and it can flow plastically. This behavior inhibits the spilling out of the grease from the bearing cages and behaves as a reservoir. The flowability of grease is influenced by its consistency and viscosity. Greases are non-Newtonian fluids and their viscosity abruptly decreases as the shear rate increases. Greases have poor oxidation stability and thermal conductivity in comparison to lubricating oil.

### 4.1.1 Current Status of the Grease Market

The major grease manufacturing companies in India are Balmer Lawrie & Corporation Ltd., Bharat Petroleum Corporation Ltd., Indian Oil Corporation Ltd., and Standard Greases & Specialties. The total production of all types of greases at the global level is ~1174 TMT, and the share of India is ~82 TMT in the year 2018 [4]. The different greases are manufactured with variation in the type of thickener at a worldwide scale, and the contributions of Indian companies are provided in Table 4.1. Approximately 72% share of the total demand is manufactured with conventional lithium and complex lithium greases.

### 4.1.2 The Need for Bio-Greases

The crude-based oil is the preferred base stock in conventional lubricating greases. Crude oil-based lubricants/greases are toxic, non-renewable, and their biodegradability is very poor [5]. The disposal of a drained-out lubricant is a very challenging task.

**TABLE 4.1**
**The Percentage Share of Different Types of Grease Production with the Type of Thickener [4]**

| Type of Thickener | World (%) | India (%) |
|---|---|---|
| Aluminum soap | 0.29 | 2.31 |
| Aluminum complex soap | 3.95 | 0.13 |
| Calcium hydrated soap | 3.02 | 3.09 |
| Calcium anhydrous soap | 4.81 | 2.72 |
| Calcium sulfonate soap | 3.68 | 1.06 |
| Calcium complex soap | 0.80 | 0.11 |
| Lithium soap | 50.94 | 73.58 |
| Lithium complex soap | 21.31 | 9.44 |
| Sodium soap | 0.42 | 2.64 |
| Other metallic soap | 0.88 | 3.69 |
| Polyurea | 6.09 | 0.21 |
| Organophilic clay | 2.29 | 0.83 |
| Other non-soap (except clay) | 1.52 | 0.20 |

Inappropriate disposal causes groundwater contamination and loss of aquatic life; this shows an adverse effect on the environment and ecosystem. Unsustainability, continuous depletion, and rising prices of petroleum products are the primary concerns. With the growing awareness about the environment, it is imperative to find an alternative base stock for lubricants.

The non-toxicity, biodegradability, renewability, and eco-friendly nature of vegetable oils enfold the attention for replacing the conventional base stock. In the last few years, several tribological investigations were conducted on various types of vegetable oils to evaluate their lubrication potential, and the results revealed that vegetable oils have excellent tribo-performance in contrast to conventional lube oils [6,7]. Vegetable oils are also used to synthesize diesel, termed 'bio-diesel,' a promising alternative to the crude fuel. Bio-diesel promotes reduced $CO_2$ emissions and global warming [8]. Vegetable oils are renewable and eco-friendly. Therefore, vegetable oil-based lubricants are also referred to as 'bio-lubricants,' 'green lubricants,' or 'eco-friendly lubricants.' The polar nature of vegetable oils confers high affinity to the metal surfaces, leading to superior lubrication performance [9]. The wide variety of vegetable oils used for different lubrication applications are listed in Table 4.2.

## 4.2 GREASE COMPOSITION

A standard grease comprises 5%–20% thickener, 80%–90% lube oil, and 0%–10% package of additives. It is a stable colloidal solution in which lubricating oil is thickened through the dispersion of a thickener. The fibrous network of the thickener forms the body structure of the grease. This network restrains the lubricating oil within the voids formed between the fibrous structures. Additives impart unique qualities to the grease, which enhance the physicochemical and tribological performance of

**TABLE 4.2**
**Various Applications of Different Vegetable Oils [5,10,11]**

| Bio-Based Oil | Applications |
|---|---|
| Canola oil | Metal working fluids, chain oil bar lubes, tractor transmission fluids, hydraulic oils, food grade lubes, penetrating oils |
| Castor oil | Gear lubricants, greases |
| Coconut oil | Gas engine oils |
| Crambe oil | Surfactants, intermediate chemicals, grease |
| Cuphea oil | Motor oil, cosmetics |
| Jojoba oil | Lubricant, cosmetics, grease |
| Linseed oil | Varnishes, stains, lacquers, paints, coating |
| Olive oil | Automotive lubricants |
| Palm oil | Grease, rolling lubricant |
| Rapeseed oil | Greases, air compressor-farm equipment, chain saw bar lubricants |
| Safflower oil | Diesel fuel, light-colored paints, enamels, resins |
| Soybean oil | Hydraulic oil, plasticizers, disinfectants, pesticides, detergents, shampoos, soaps, coating, paints, printing inks, metal casting/working, bio-diesel fuel, lubricants |
| Tallow oil | Soaps, steam cylinder oils, cosmetics, plastics, lubricants |

the grease. The lubrication performance and physical characteristics of greases are dependent on the properties of their ingredients.

## 4.2.1 Base Oil

The base oil is a hydrocarbon compound, extracted from biological and non-biological sources. The natural oil derived from the crude oil is under the non-biological oil category; if it is extracted from plant seeds, it falls into the biological group. Base oils are classified as mineral, synthetic, and vegetable oils. The physical properties such as viscosity, viscosity index, oxidation stability, volatility, polarity, and solvency affect the tribological performance of lubricants/greases.

### 4.2.1.1 Mineral Oil

Crude oil is a complex mixture of straight and branched kinds of paraffin, naphthenes, aromatic hydrocarbons, aromatic heterocyclics, alkenes, waxes, and so on. Mineral oil is obtained from crude oil through a fractional distillation process. They are preferred as a base stock in the formulation of a wide variety of lubricants and greases due to their low cost, abundant resource, and enormous viscosity grades. Based on the chemical composition of mineral oils, they are further classified into paraffinic, naphthenic, and aromatic oils, and their structures are shown in Figure 4.1.

#### *4.2.1.1.1 Paraffinic Oil*

Paraffinic oil has either a straight-chained or branch-chained hydrocarbon structure. It is an extensively used category of mineral oil as a base stock in the formulation of lubricants/greases. This oil is non-polar and shows higher viscosity at low

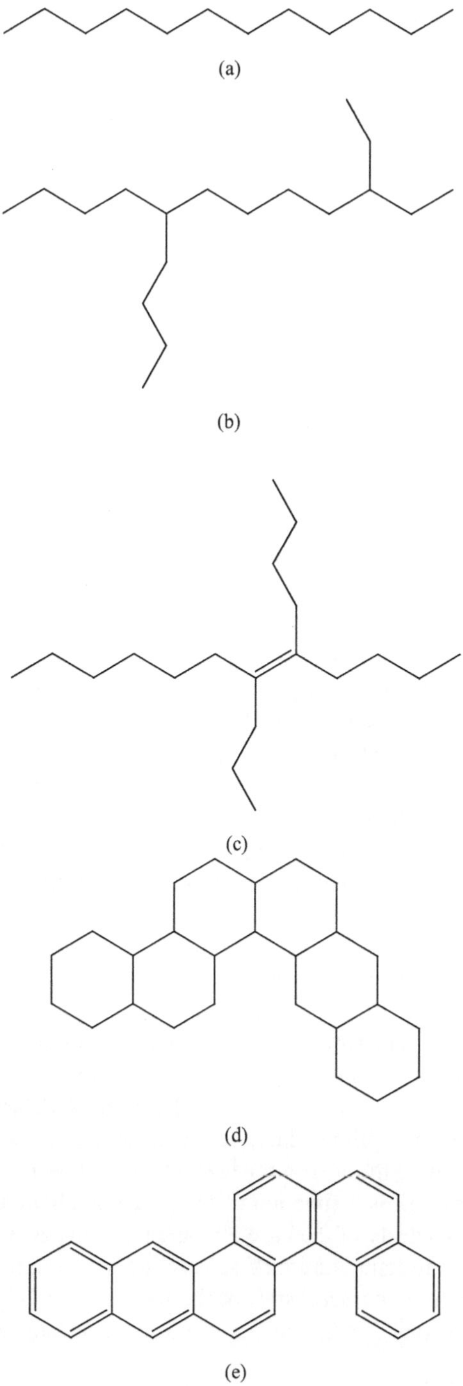


**FIGURE 4.1** Structures of various mineral oils: (a) straight-chained paraffinic, (b) branch-chained paraffinic, (c) olefinic, (d) naphthenic, and (e) aromatic mineral oils.

temperatures, attributed to the presence of wax. The hydrocarbon includes carbon–carbon double bonds that are unsaturated and commonly known as olefins. Olefinic oil has poor oxidation stability and a higher pour point because of the presence of unsaturated hydrocarbons. Paraffinic oils are poor solvents than olefinic oils. If the number of branch-chained hydrocarbon increases in paraffinic oil, it adversely affects the viscosity index (VI) and the melting point of the oil.

##### *4.2.1.1.2 Naphthenic Oil*

Naphthenic oil consists of six carbon atoms (sometimes five) in the cyclic ring called cycloalkanes. It may contain one or more cyclic rings, and each ring has a saturated carbon chain. This oil is free from wax and has high densities as well as low pour points in comparison to straight-chained hydrocarbons. Naphthenic oil has excellent solvency for grease thickeners and is more polar than paraffinic oils. The metallic soap is less soluble in paraffinic oils than naphthenic oils. This merit of naphthenic oil makes it the most favorable base stock for grease. Naphthenic oils have low-temperature flowability and a low viscosity index. At elevated temperature applications, this oil demonstrates poor viscosity–temperature characteristics and oxidation stability. Paraffinic oil showed higher VI, pour point, and oxidation stability than naphthenic oil.

##### *4.2.1.1.3 Aromatic Oil*

Aromatic oil has six carbon atoms arranged in a cyclic structure having alternate double and single bonds. This oil possesses excellent thermal stability, so it has widespread use in heat transfer fluids. Aromatic oils show high density and poor viscosity–temperature characteristics. Aromatic compounds are naturally free from wax; they have low melting points and low-temperature fluidity. Aromatics also have high polarity. Therefore, they have a strong tendency to dissolve in water. Aromatic compounds are excellent solvents; consequently, they are widely used in rubber compounding. In comparison with paraffinic and naphthenic oils, aromatic compounds have superior thermal stability.

The properties of a base oil depend on its molecular weight distribution and chemical structure (see Table 4.3). At high temperatures, the volatility of a base oil plays a crucial role, which affects the consumption of lubricants, particularly if the base oil has a low viscosity grade. The solvency (capability to dissolve additives) of the base oil depends on its chemical structure. A hydrocarbon with ring structures interacts better with additives in contrast to straight-chained hydrocarbons. Polarity defines the interaction ability of a base oil with available surfaces and additives. Therefore, polarity plays a vital role in forming the adsorption layer of base oil over the lubricated surface. The viscosity of a base oil is a function of hydrocarbon chain length. As the chain length increases, the viscosity of the base oil increases. Viscosity of a base oil is very sensitive to the operating temperature. Viscosity index is a parameter that indicates the viscosity–temperature characteristics of the base oil. The higher viscosity index of the base oil suggests that the variation in viscosity at elevated temperatures is low.

#### **4.2.1.2 Synthetic Oil**

Synthetic oils are synthesized from one or more base oils to acquire the desired properties. Generally, synthetic oils are made from a petroleum base stock. The cost of

**TABLE 4.3**
**Some Properties of Different Hydrocarbon Mineral Oils [12,13]**

| Property | Paraffinic | Naphthenic | Aromatic |
|---|---|---|---|
| Pour point | High | Low | Low |
| Oxidation stability | High | High | Low |
| Density | Low | Low | High |
| Viscosity index | High | Low | Low |
| Volatility | Low | Medium | High |
| Thermal stability | Low | Low | High |
| Toxicity | Low | Low | Medium |

synthetic lube is much higher than crude-derived lubricants. Polyalphaolefins (PAO), polybutylene, polyisobutylene, polyalkylene glycols (PAG), polyphenyl ethers, perfluoropolyalkylether (PFPE), polyol esters, phthalate esters, synthetic polymeric esters, phosphate esters, synthetic hydrocarbon oils, silicones, and polysiloxanes are common examples of synthetic base stocks. Synthetic oils have a higher viscosity index, lower evaporation losses, high oxidation resistance, and excellent low-temperature fluidity than mineral oils. Synthetic oils are not compatible with conventional thickeners so fumed silica, polyurea, clay, and Teflon are widely used as a thickener in the development of synthetic oil-based greases. Synthetic greases are applied where mineral oil-based greases do not meet the desired service level.

#### 4.2.1.3 Vegetable Oil

The crushing of kernels of plants provides vegetable oils. There are two types of vegetable oils: edible and non-edible oil. In a comparison of petrolubricants, vegetable oils possess required lubrication properties, as listed in Table 4.4. However, vegetable oils have a significant limitation for their poor oxidative and pour point properties. Vegetable oils are esters composed of a triglyceride molecular structure consisting of a long hydrocarbon chain of fatty acids linked with glycerol. If a hydrocarbon chain combines with the COOH (carboxyl) group at one end, then it is known as 'fatty acid (FA).' The hydrocarbon chain length of FA varies from a range between 4 and 36 carbon atoms [7]. Carbon–carbon single bonds are more chemically stable than carbon–carbon double or triple bonds. If there is no double bond in the FA hydrocarbon chain, then it is 'saturated' FA. If the hydrocarbon FA chain contains only one double bond, it is termed 'monounsaturated' FA. If the count of double bonds is more than one in the FA hydrocarbon chain, then it is referred to as 'polyunsaturated' FA. Vegetable oils are composed of saturated, monounsaturated, and polyunsaturated FAs. As the number of double bonds increases in the FA chain length, oil shows poor oxidation stability and higher liquidity. Various chemical treatments such as epoxidation [14,15], transesterification [16], and hydrogenation [17] are available to improve the limitations such as low-temperature fluidity and oxidation stability of vegetable oils. Saturated vegetable oils have lower liquidity (high melting point) and more stability against oxidation [18].

**TABLE 4.4**
**Comparison between Physicochemical Properties of Mineral Oils and Vegetable Oils [19]**

| Properties | Mineral Oils | Vegetable Oils |
|---|---|---|
| Biodegradability (%) | 10–30 | 80–100 |
| Cold flow behavior | Good | Poor |
| Density (@ 20°C, kg/m$^3$) | 840–920 | 890–970 |
| Evaporation loss | Higher | Lower |
| Flash point, °C | Lower | Higher |
| Fire point, °C | Lower | Higher |
| Hydrolytic stability | Good | Poor |
| Load-bearing capacity | Low | High |
| Oxidation stability | Good | Fair |
| Pour point, °C | −15 | −22 to +12 |
| Shear stability | More than vegetable oil | Minimum |
| Toxicity | Toxic | Non-toxic |
| Viscosity index | 100 | 100–200 |
| Volatility | High | Low |

### 4.2.2 Thickener

Thickener is one of the critical ingredients in grease that governs most of the physiochemical properties of grease and grease is categorized based on the type of thickener used. Thickener controls the consistency of grease, water washout characteristics, and so on. A wide variety of thickeners are available for greases. Thickeners are generally of two types, i.e. soap and non-soap. Grease classification is based on the type of thickener because it provides preliminary information about grease quality, selection of grease, and its application. Apart from this, greases are also classified based on their end-use. For example, sealing grease is used to prevent leakage or clearance between mating surfaces, grouped into the vacuum, thread, and valve greases. Antifriction grease protects the interacting surfaces against friction and wear and is split into general-purpose grease (average temperatures) and multipurpose grease (elevated temperatures).

#### 4.2.2.1 Soap Thickener

Salt is a chemically neutralized product of an acid and base, and water is obtained as a by-product. When FA is neutralized with strong base, the reaction product is referred to as 'soap.' Soaps are further divided into four groups: alkaline-earth-metal (barium and calcium), alkali metal (lithium and sodium), heavy metals (zinc and lead), and aluminum soap. Furthermore, each group of soap is categorized into subgroups of straight and complex. Complexing agents are organic or inorganic materials added in the conventional metallic soap for the development of complex grease. Metallic acetate, chloride, benzoate, and carbonate are typical examples of complexing agents. The primary purpose of adding a complexing agent in the grease is to enhance the dropping point. The dropping point of complexing agent thickened grease is 50°C–100°C higher than conventional soap thickened grease [20]. Conventional and complex lithium soap-based greases are preferred globally, approximately 70% greases of total demand worldwide [18].

Shorter and longer hydrocarbon chain lengths degrade the thickening capacity. The 12-hydroxystearic acid (having 18 carbon atoms) derived from vegetable oils show a maximum thickening effect [20]. The solubility of a thickener in the base oil increases with increasing the carbon chain length or vice-versa. The molecular weight of the base oil is also a significant factor that affects the solubility of thickener. Unsaturated carboxylic acids are more soluble in mineral oils, lower the drop point, and decrease the thickening capacity. Their applications are limited due to their poor oxidation stability [20]. The presence of a branched alkyl chain reduces the thickening capacity and melting point of a thickener. On the other hand, hydroxyl groups have a higher polarity, which increases the thickening capacity and melting point of the thickener [21].

#### 4.2.2.2 Non-soap Thickener

Non-soap greases are organic or inorganic materials. Polyurea and PTFE are typical examples of organic non-soap having spherical or plate-like fine particles. Inorganic non-soaps are very fine particles having enough surface area and porosity to absorb oil; clay and silica are typical examples of inorganic non-soaps. These non-soaps do not show any phase change or melting point at elevated temperatures. Therefore, these greases are preferred for high-temperature applications.

#### 4.2.2.3 Mixed Thickener

If a grease contains more than one soap cation, it is referred to as 'mixed-soap-grease.' Lithium-calcium (LiCa), calcium aluminum (CaAl), and calcium-sodium (CaNa) are typical examples of mixed greases. The terminology of mixed grease comprises the name of both the soap cations; the portion which is higher in the mixture is designated first, and the portion which is lower is designated next [22]. For example, lithium-calcium mixed grease indicates that the portion of lithium cations is higher than calcium cations in the mixture, so lithium comes first in the nomenclature followed by calcium. The properties of mixed greases are distinct from their parent grease. Lithium-calcium grease demonstrates better shear stability and water resistance in comparison to raw lithium grease. The 20% w/w concentration of calcium soap in lithium soap increases the dropping point and improves the antifriction and antiwear performance of mixed grease compared to raw lithium grease [20]. Similarly, a minor proportion of clay thickener mixed with aluminum complex grease results in improved mechanical properties, and heat resistance characteristics achieved are better than pure aluminum complex grease [23].

## 4.3 GREASE STRUCTURE

Soap-based greases have a fibrous structure, which is shown in Figure 4.2. The interlocking between the fibers of the thickener occurs through weak van der Walls forces

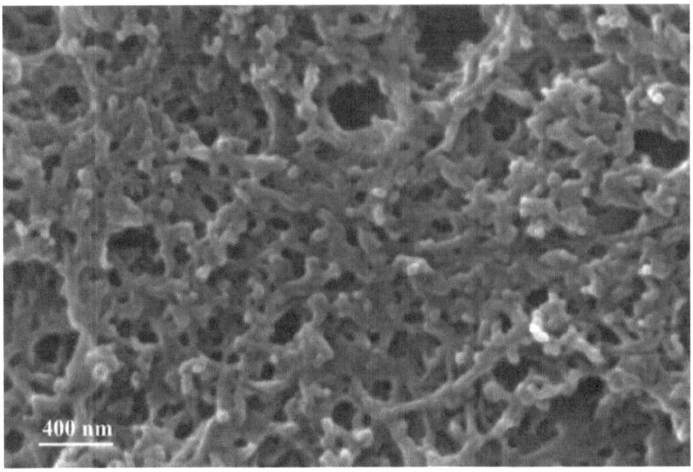


**FIGURE 4.2** SEM micrographs of a 12-lithium hydroxy stearate soap network after extraction of base oil.

or ionic and dipole–dipole interactions, including hydrogen bonding [24]. The interaction of fibers with each other affects the effectiveness of the forces, as mentioned earlier. These interlockings of fibers form voids in the order of $10^{-6}$m [25]. The base oil is trapped within the voids created by the fibrous network of the grease. In other words, the grease fibrous network impersonates a reservoir of base oil. The base oil interacts with the fibers with a combined effect of capillary and van der Walls forces [24]. Salomonsson et al. have found that the average length and diameter of fibers of lithium grease were 1 μm and 30 nm, respectively [26]. The length of soap fibers is a crucial factor that affects the lubrication performance of the grease [27]. Non-soap-based greases are not fibrous structures. Therefore, to obtain the same level of consistency, non-soap-based greases require a higher concentration of thickener than soap-based greases [28]. The thermal aging of grease affects the structure of the fibers. Couronne et al. have observed that thermal aging showed an adverse effect on the fibrous structure of lithium and lithium complex greases while a prolific effect with di- and tetra-urea greases [29].

## 4.4 ADDITIVES

The period of usefulness of greases or lubricants is limited. Various factors, such as severe loading conditions, high temperature, high speed, environmental conditions, and mechanical and thermal stresses, affect the lifetime of grease. Therefore, grease does not achieve the performance measures for which it is designed. As grease ages, it becomes dry and brittle, so it does not exhibit lubrication performance satisfactorily. To extend the useful lifetime and to improve the lubrication performance of grease, there is a need to add some additives and fillers. A typical grease usually contains a 0%–10% share of additives, depending on the application. The prime objective of additives is to enhance the physicochemical and tribological properties of the greases.

**TABLE 4.5**
**Common Examples of Various Kinds of Additives Used in Greases [21,22,32]**

| Additives | Examples |
|---|---|
| Antifriction and antiwear | Zinc dialkyl dithiophosphate, metal oxides, carbon-based materials, nanomaterials |
| Corrosion inhibitors | Metal sulfonates (sodium, barium, calcium, ammonium); metal naphthenates (zinc, lead, bismuth); metal carboxylates (zinc, calcium); alkenyl succinic acid esters; sarcosine derivatives; imidazoline; amides; phosphate esters, and so on |
| Extreme pressure | Phosphate and thiophosphate esters; sulfurized fats, olefins, and esters; sulfides and disulfide; metal dialkyl dithiophosphate (zinc, antimony); graphite |
| Oxidation inhibitors | Hindered phenols; metal dialkyldithiophosphates; aromatic amines; metal dialkyldithiocarbamates; zinc dialkyl-dithiophosphates; diaryldisulfides; and so on |
| Tackiness | Latex compounds; ethylene–propylene olefin copolymer (OCP); polyisobutylene; copolymer; polybutene; and so on |

The additives utilized in the lubricants (liquid) are almost the same for grease applications. The addition of additives in the grease modifies its tribological and rheological performance. Some additives destroy or alter the fibrous microstructure of the grease [30,31], which directly affects the lubrication performance. Industrial greases are comprised of a basic package of additives such as corrosion inhibitors, oxidation inhibitors, antifriction (AF), antiwear (AW), extreme pressure (EP), and tackiness additives, viscosity index improvers, and wear protection additives. If grease is used under severe conditions, then grease is enriched with an enhanced package of additives. Examples of widely used additives are mentioned in Table 4.5.

Corrosion inhibitors are incorporated to prohibit the electrochemical oxidation of the metallic surface and their dissolution. They are sorted as anodic and cathodic inhibitors. Anodic inhibitors inhibit the anodic reaction of metals and their dissolution, while cathodic inhibitors prevent cathodic reaction that results in decreasing the formation of oxygen or hydrogen ions. Oxidation inhibitors are used in grease to minimize the reaction rate of hydrocarbons with oxygen and prevent the formation of hydroxides and peroxides. AF/AW additives are used to minimize the scuffing and wear of mating surfaces, and EP additives are added to enhance the load-bearing capacity of the grease under severe loading conditions. Grease is doped with tackiness additives, improves cohesive and adhesion properties, increases water wash-off resistance, and throws off resistance from bearings. Tackiness additives also provide extra cushioning and reduce noise.

## 4.5 ROLE OF NANOADDITIVES

In the past two decades, nanomaterials have been used as additives in grease lubrication, and their performance has immensely attracted lubrication engineers. The

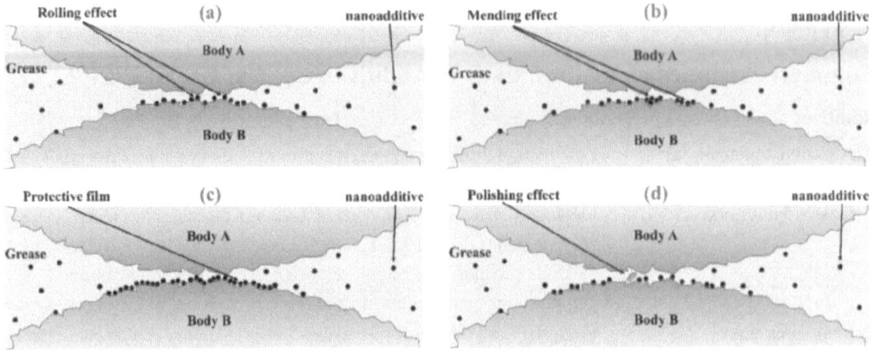


**FIGURE 4.3** Schematic representation demonstrating the role of nanoadditives in grease lubrication: (a) rolling effect, (b) mending effect, (c) protective film, and (d) polishing effect.

effectiveness of nanomaterials is dependent on the morphology, size, concentration, and compatibility with the base stock. They quickly enter at the tribopairs due to their nanosize and prevent direct asperities contact. Nanomaterials have different morphologies, viz. nanosheets [33], nanotubes [34], nanospheres [35], nanohorns [36], etc., which have a significant influence on the tribological performance of the grease. Enormous investigations have been carried out to find the optimum concentration of nanomaterials in lubricants, and it is a critical factor that affects the tribological performance [37–39]. If the concentration of nanomaterials is less or more than its optimum value, it is detrimental to friction reduction and wear protection ability [39].

Various lubrication mechanisms have been proposed by the researchers to explain the role of nanomaterials. These mechanisms include protective film formation [33,40,41], polishing effect [35,42,43], ball-bearing effect [42,44], and mending effect [35,43,44]. A schematic representation of the abovementioned mechanism is depicted in Figure 4.3.

Spherical nanomaterials are believed to behave like nanoball-bearings that roll between the rubbing surfaces. The sliding friction converts into a combined effect of sliding and rolling due to the presence of spherical nanomaterials. A schematic illustration of the rolling effect is shown in Figure 4.3a. Figure 4.3b shows the schematic image of the mending effect. The mending effect is also termed the 'self-repairing effect.' In this phenomenon, nanomaterials have filled the scars and grooves of the worn surfaces. The schematic representation of a protective film is depicted in Figure 4.3c. During this phenomenon, a tribo-film or absorption film was developed on the interacting surfaces. This film was believed to protect the rubbing surfaces, which leads to minimizing wear and friction. X-ray photoelectron spectroscopy (XPS) or Raman spectroscopy analysis techniques were used to verify the formation of tribofilms. The polishing effect is explicitly being illustrated in Figure 4.3d. The polishing effect is also referred to as a 'smoothing effect.' Nanomaterials assist in the reduction of abrasion of the interacting surfaces. This smoothing effect is believed to improve the tribological characteristics of lubricants, mainly attributed to the reduction of surface roughness.

## 4.6 BASIC PROCESS FOR GREASE PREPARATION

The grease formulation procedure is primarily based on the nature of the thickener. The various grease synthesis procedures were reported, showing a great diversity [30,35,45–49]. Typically, the preformed soap is commercially available in the form of powder or flakes. The conventional procedure of grease synthesis consists of dispersion of a stoichiometric amount of preformed soap in the base stock at a high temperature under defined conditions and stirring of the dispersed solution continuously. The consistency of grease is controlled with the addition of the remainder of base oil. The blending of additives enhances physicochemical properties. In the final phase, the grease is subjected to mechanical working to homogenize the composition. This technique is preferred only for the synthesis of synthetic grease due to the higher cost of preformed soap [20].

The soap-based grease formulated through the in-situ scheme, a stoichiometric amount of FA, and alkali are added in the base oil. The dispersion mixture is heated and stirred continuously under defined conditions. In this way, the soap is formed in-situ through the saponification reaction between FA and alkali. It is necessary to evaporate the water produced during the reaction and allow the soap to crystallize. Furthermore, the remaining base oil is added to control the consistency followed by the blending of additives. In the final step, the grease is milled thoroughly to obtain a homogeneous finished product. Sometimes, several mixing cycles are required to achieve the desired properties of the grease.

Soap-based greases have some limitations when exposed to extremely hot or cold temperatures. Non-soap-based greases show an excellent lubrication performance at extreme hot and cold temperatures than conventional soap-based greases. Usually, the non-soap-based thickener is mixed with a base oil at room temperature to form a thickened product like conventional greases. However, some non-soap greases (viz. polyurea and polyurea-complex) require heat and chemical reactions in a similar fashion as soap-based greases. These chemical reactions are not like a saponification reaction of FA and alkali. Here, amines and isocyanates react with each other to form urea. The oxidative stability of polyurea grease is exceptionally high because of the absence of acids and metals. In the synthesis of organoclay-based grease, an organoclay thickener along with acetone (activator) is mixed with a base oil by a high shear milling process. After completion of the process, acetone can be evaporated. Typically, organoclay greases do not have a dropping point, so these greases are widely used under high-temperature conditions such as bakery ovens.

Overall, in the grease synthesis process, one of the most significant concerns is the blending of additives. Generally, soap-based greases are heated at high temperatures depending on the type of thickener. Some additives are highly sensitive to heat and become ineffective at such high temperatures. Therefore, conventionally, the blending of additives is preferred when the greases cool down at less than 85°C. If additives are highly resistive against temperature or remain unaffected at high temperatures, they can be blended at any time in the grease.

It is mandatory to control all process variables to produce a perfect finished grease containing the desired properties. An imperfect saponification reaction increases the acidity or alkalinity of the grease, which reduces the thickening effect of the soap and shows an adverse effect on various performance characteristics. If the concentration of the thickener is significantly increased, it leads to an increase in the apparent viscosity, consistency, and shear stability and lowers the oil separation from the grease under pressure.

## 4.7 GREASE CONSISTENCY

The essential characteristic of grease is consistency. The word 'consistency' indicates the thinness or thickness related to grease flowability. Primarily, the consistency of grease is mainly controlled through the type and concentration of thickener, but the viscosity of base oil has a minor influence on it. The consistency of grease is measured using the cone penetrometer as per the ASTM D 217 standard [50]. A schematic representation of the cone penetration test is shown in Figure 4.4. When the quantity of the grease sample is limited, then one-half (1/2) and one-quarter (1/4) scale cone equipment is used to determine the consistency of the grease as per the ASTM D 1403 standard [51]. The prescribed test conditions are summarized in Table 4.6. This test is aimed to evaluate the unworked and worked penetration depth, which designates the NLGI (National Lubricating Grease Institute) number to the grease. In this test, a standard cone is released to sink freely in the grease for 5 seconds at 25°C. The penetration depth is measured by a dial gauge, attached with the cone penetration equipment. The unit of measured penetration is in tenths of a millimeter. The test conditions for unworked and worked penetration measurements are the same as described in Table 4.6. If the penetration measurements are acquired

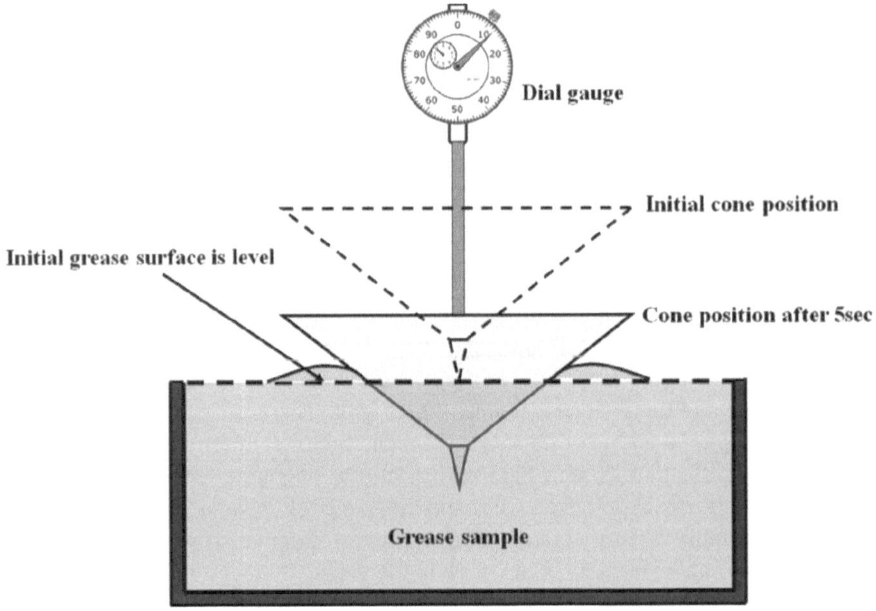


**FIGURE 4.4** Schematic image of a cone penetration test setup. (Adapted from Ref. [39] with permission.)

**TABLE 4.6**
**Prescribed Test Conditions Used to Determine the Grease Consistency**

| ASTM Standard | Type of Scale | Weight of Cone Assembly (g) | Quantity of Grease (g) | Duration (sec) | Temperature (°C) |
|---|---|---|---|---|---|
| D 217 | Full scale | 150 ± 0.100 | 400 | 5 ± 0.1 | 25 ± 0.5 |
| D 1403 | Half scale | 37.5 ± 0.050 | – | 5 ± 0.1 | 25 ± 0.5 |
| | Quarter scale | 9.38 ± 0.025 | – | 5 ± 0.1 | 25 ± 0.5 |

with 1/2 scale or 1/4 scale, the cone penetrometer needs to be converted to full scale as per the following equations [51].

For 1/2 scale,

$$P = 2r + 5 \quad (4.1)$$

where

$P$ = cone penetration by Test Method D 217 and
$r$ = cone penetration by 1/2 scale equipment

For 1/4 scale,

$$P = 3.75p + 24 \quad (4.2)$$

where

$P$ = cone penetration by Test Method D 217 and
$p$ = cone penetration by 1/4 scale equipment

In unworked penetration, the measurements are obtained on the undisturbed grease structure. In worked penetration, a grease worker is used to shear the grease structure. The image of a 1/4 scale grease worker is depicted in Figure 4.5. The grease worker has a perforated disk, which shears the grease structure through the holes twice at every stroke of the grease worker's arm. As per the standard 60 strokes, furnished grease is used to measure the worked penetration depth. The mechanical working amends the consistency of the grease. The NLGI number designation of the grease is based on worked penetration depth measurements. NLGI number classifies the grease's hardness in the range between 85 (NLGI 6) and 475 (NLGI 000). NLGI 000 attributes to 'fluid,' and NLGI 6 ascribes for 'hard' grease. The classification system of NLGI is depicted in Figure 4.6.

## 4.8 DROPPING POINT

The dropping point is another significant characteristic of grease. The dropping point is a temperature at which grease transforms its phase from the semi-solid phase to liquid. The dropping point is referred to as the 'melting point' of the grease. At this

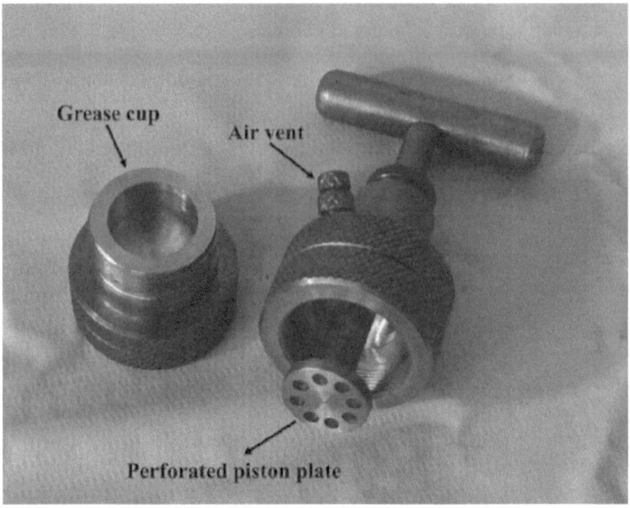


FIGURE 4.5 Image of a 1/4 scale grease worker. (Adapted from Ref. [39] with permission.)

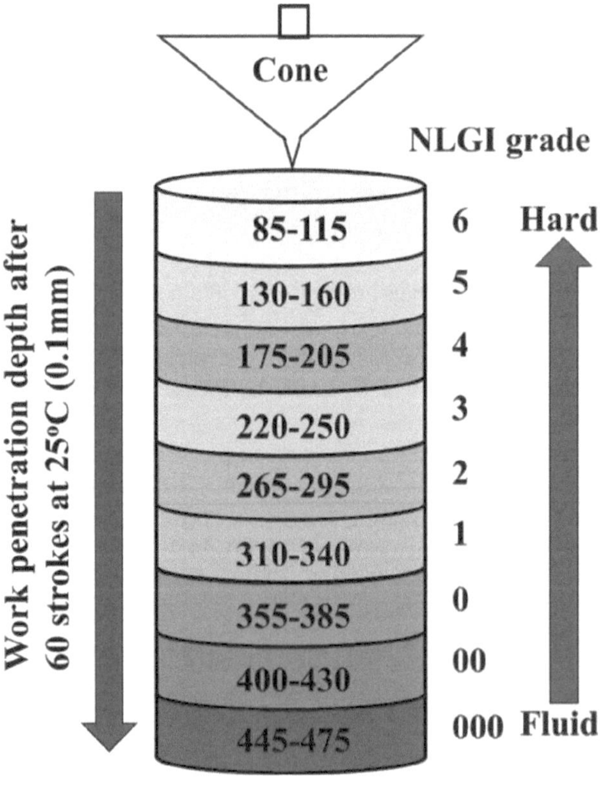


FIGURE 4.6 The classification of NLGI number.

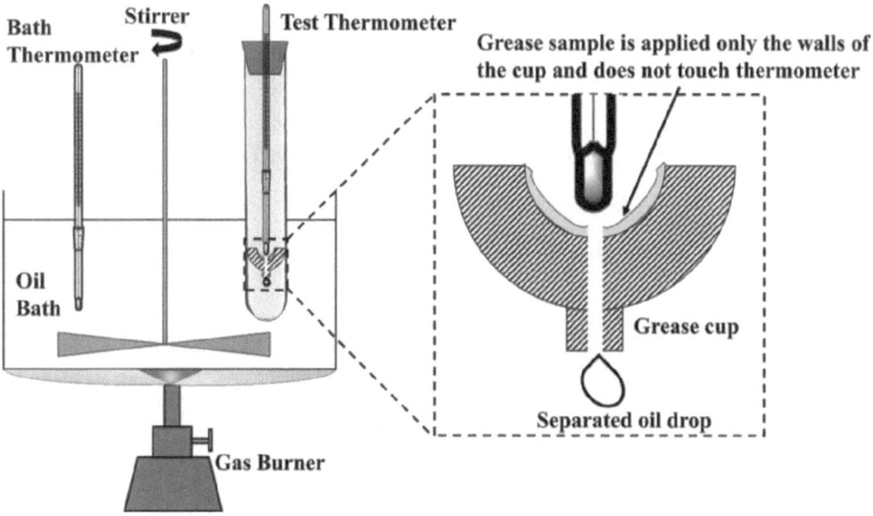


**FIGURE 4.7** Schematic representation of dropping point equipment. (Adapted from Ref. [39] with permission.)

temperature, the thickener's property fades to constrain the base oil within its fibrous network. Some non-soap-based greases do not melt with the increment of temperature and can be stable up to the decomposition temperature of either the thickener or the base oil. Therefore, these greases do not show a drop point and are preferably used for bakery ovens. The schematic illustration of dropping point equipment is displayed in Figure 4.7. In this test, the grease is heated under prescribed conditions as per the ASTM D 566 standard [52]. When the first drop of base oil detaches from the grease matrix and falls in the bottom of the tube, that temperature is referred to as the 'dropping point,' which is an average of two thermometers reading. In this test method, the bath temperature is limited up to 288°C and the dropping point of grease is determined over the wide range of temperature through the ASTM D 2265 standard [53]. The thickener used in the formulation of the grease affects not only the dropping point but also the maximum useful working temperature of the grease. The effect of thickener on the dropping point of the grease is represented in Figure 4.8. The standard deviation indicates the range of variation in the dropping point of grease. The PTFE and lithium complex thickeners show the highest dropping point among various thickeners, and their maximum working temperature ranges up to 177°C.

## 4.9 BIO-BASED GREASES

Typically, greases are manufactured using mineral oils, synthetic esters, and vegetable oils, and their shares in global demand are 90%, 9%, and 1%, respectively [54]. Environmental concern and rapid depletion of crude oil emphasize finding a substitute for the conventional base stock. The physicochemical properties of vegetable

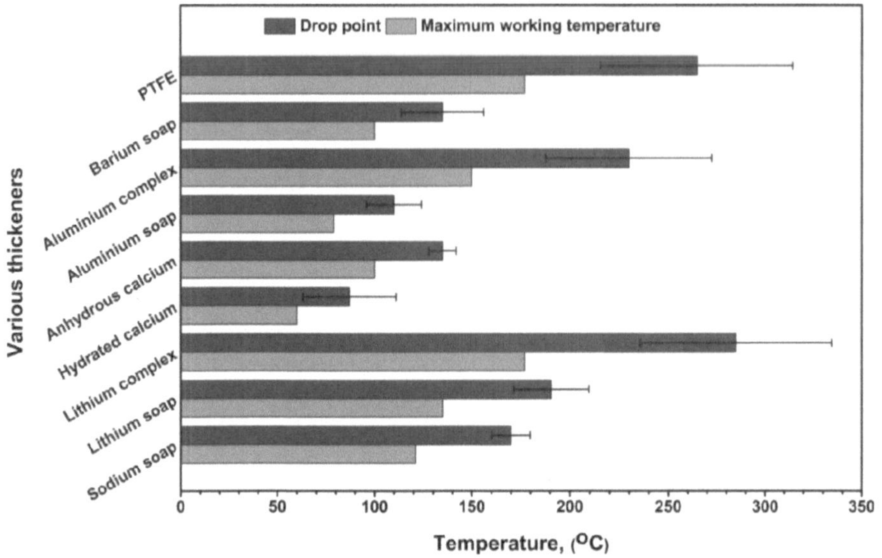


**FIGURE 4.8** Effect of a thickener on the dropping point of the grease.

oils are already described in 'Section 4.2.1.3,' and they are suitable candidates to substitute conventional petroleum-based oil. It is believed that bio-based grease can reduce environmental pollution due to its excellent biodegradability. The degree of biodegradability of grease is based on ingredients used in the composition. The biodegradability of different oils is reported, such as mineral oils (20%–40%), vegetable oils (90%–98%), and synthetic esters (65%–100%) [55].

The FA (vegetable oil) hydrocarbon chain linked with glycerol and its presence in the grease increase its affinity with water and deteriorate its flowing tendency at cold temperatures. The drawbacks of vegetable oils (viz. pour point and oxidation stability) have been rectified with various chemical treatments. Chemically modified vegetable oils are believed to be the most promising alternative for mineral-based oils in the development of conventional lubricating grease. It means that soap-based grease with vegetable oils can be formulated as described in 'Section 4.6'. Honary et al. reported that soybean oil-based greases were commercialized in the United Nations to lubricate rail curves and flanges of locomotive wheels. Furthermore, bio-based greases have applications in the greasing of the truck fifth wheel [18].

In general, the presence of unsaturation in vegetable oils made it significantly prone to oxidation when compared with mineral oils. Oxidation degradation is not a significant concern if base oil is entrapped within the fibrous mesh of the thickener as long as possible. When base oil is discharged from the fibrous network of the thickener under pressure or due to degradation of the fibrous structure, oxidation degradation of base oil becomes more predominant. In this context, some investigations on soybean oil-based greases were carried out to comprehend the effect of chain length and composition on the thermo-oxidative behavior of the grease [56–58]. Suitable antioxidants in optimum concentration can provide excellent oxidation stability to the soybean oil-based grease [57]. Besides, a high soap-to-oil ratio (range from 1:2 to 1:4) has improved the lubricity and oxidation stability of the grease. Adhvaryu et al. [56] have inferred from their study that the chain length of FA used in lithium soap does not significantly affect the oxidation and thermal stabilities of the grease.

## 4.10 EVALUATION OF TRIBOLOGICAL AND RHEOLOGICAL PROPERTIES OF BIO-GREASES

The interacting surfaces encounter friction and wear. Many engineering components demand less friction and wear between tribopairs. High friction leads to deterioration of the efficiency, reliability, and life of the machine. Therefore, lubricants/greases protect the tribopairs against friction and wear, improve efficiency, and prolong life of the machine components. The evaluation of the lubrication performance of the grease is referred to as the 'tribological study' of the lubricant/grease. The grease not only exhibits solid-like behavior but also demonstrates liquid-like behavior. In other words, grease is a 'visco-elastic' material. The evaluation of the flow behavior of the grease is entitled as the 'rheological study' of the grease. The tribological and rheological studies of lubricating greases are essential to understand the exact lubrication mechanism and performance of the greases.

### 4.10.1 Tribological Properties of Bio-Greases

The load-bearing capacity and capability of the greases to minimize friction and wear are assessed using tribometers. Various tribometers such as four-ball tester, pin-on-disk, FZG gear tester, SRV tribometer, ball-on-disk, and so on are available to evaluate the tribological performance of the greases. The tribological performance of the grease is assessed with a different type of base oil and its viscosity, type of thickener and its concentration, and the package of additives. Several research articles are available on the tribological performance of conventional greases [42,43,48,49,59,60]. Here we focus on the tribological investigations of vegetable oil-based greases.

According to the current scenario, the biodegradability of a lubricant is a crucial facet, and the biodegradability of vegetable oil-based greases is superior to that of conventional greases. Florea and coworkers [55] have developed biodegradable greases with soybean oil, rapeseed oil, and castor oil, which were selected as the base oil, and lithium soap was used as a thickener. In addition, another biodegradable grease was formulated with synthetic ester (dioctyl adipate and dioctyl sebacate) as a base oil and organo-clay as a gelling agent. The biodegradability test (CEC-L-33-A-94 method) results exhibited the biodegradability of vegetable oil-based grease, and diester greases are over 85% and 80%, respectively. Furthermore, $\alpha$-tocopherol was used as an antioxidant additive in biodegradable grease and it improved the oxidation stability. These results indicate that vegetable oil-based greases have superior biodegradability than synthetic ester-based greases. The chain length of FAs and the degree of unsaturation greatly influence the lubrication performance of bio-greases. Sharma et al. [57] have studied the tribological and thermo-oxidation behavior of soybean oil thickened with lithium soap. In addition, the variable composition of thickener was used to understand the effect of FAs chain length and degree of unsaturation on the lubrication performance of the soy-grease, and the results revealed that the degree of unsaturation in FA chains decreases, consequently increasing lubricity and oxidation stability of the grease.

Barriga et al. have investigated sunflower oil-based grease thickened with polymer for heavy-duty application. It showed superior tribological performance in gear simulation test rig compared with mineral grease and relatively identical performance with slightly higher wear on a four-ball tester [61]. Similarly, epoxy soy oil-based lithium grease exhibited excellent oxidation stability and tribological behavior compared to commercially available mineral oil-based greases [58]. Refined bleach deodorized palm oil (RBDPO) and epoxy RBDPO were thickened with lithium and calcium soaps to develop additive-free palm grease. It showed better tribological performance as compared with mineral (HVI 160S) grease [62,63]. Panchal et al. have formulated bio-based grease with transesterified Karanja oil and lithium soap, which exhibited similar tribological behavior as mineral greases [54]. The jatropha vegetable oil-based lithium grease with a multifunctional additive showed excellent tribological performance compared to commercial grease [64]. The tribological results of vegetable oil thickened with metallic soap additive-free bio-greases have demonstrated superior antiwear and antifriction characteristics compared to conventional greases. Furthermore, it is concluded that the tribological performance of bio-grease can also be enhanced by adding appropriate nanoadditives.

Furthermore, lubrication engineers are endeavoring to replace traditional thickeners with environmentally friendly thickening agents. Biopolymers are gaining more attention due to their eco-friendly nature and high biodegradability. Therefore, biopolymers (viz. methyl cellulose, ethyl cellulose, Kraft cellulose pulp, and chitosan) were investigated to substitute for traditional thickening agents [65–67]. Gallego and coworkers have developed a biodegradable grease with castor oil and chemically modified biopolymers (cellulosic pulp, chitin, and methylcellulose) and evaluated for tribological performance against conventional calcium and lithium-based greases. The methylcellulose-based grease has established minimum COF with time as compared with all greases [68]. The bio-grease formulated with a combination of high-oleic sunflower oil (HOSO) and castor oil as a base oil and sorbitan monostearate as a thickening agent has demonstrated a promising tribological performance [69]. The bio-grease formulated with the dispersion of epoxide-functionalized alkali lignins (EALs) consisting of a thickener concentration of 5 wt% and epoxy indices (0.79 mol/kg) in castor oil exhibited excellent tribological performance under the mixed/EHD lubrication regime [70]. From the studies mentioned above, vegetable oil thickened with biopolymers showed its significant potential to protect the tribopairs against friction and wear. The efforts of researchers suggest that vegetable oil-based greases are biodegradable, having excellent tribological performance, and will be a suitable substitution for conventional greases (Table 4.7).

### 4.10.2 Rheological Properties of Bio-Greases

The grease flow characteristics are crucial because greases confront severe conditions such as high speed, high shear rate, and high load. Lubricating greases are

**TABLE 4.7**
**Tribological Performance of Bio-Based Greases**

| Base Oil | Thickener | Nanoparticles | Tribometer | Ref. |
|---|---|---|---|---|
| Sunflower | Polymer | – | Four-ball tester and gear simulator | [61] |
| Soybean | Lithium soap | Dithiocarbamate | Ball-on-disk | [57] |
| Soybean | Lithium soap | Antioxidants (alkylated phenol, dithiophosphoric acid ester, and diphenylamine)<br>Antiwear and load-carrying additives (1,3,4-thiadiazole, methylene-bis-dibutyldithiocarbamate, molybdenum dibutyldithiocarbamate, and a synergistic mixture of antimony dithiocarbamate/sulfurized olefin | Four-ball tester | [58] |
| Palm | Lithium soap | – | Four-ball tester and gear wear tester | [62] |
| Palm | Calcium soap | – | Four-ball tester and gear wear tester | [63] |
| High-oleic sunflower oil (HOSO)<br>Octyldodecyl isostearate (OCT)<br>Trimethylolpropane trioleate (TMPO), PAO | Highly dispersed silica (HDS)<br>Lithium soap<br>Calcium soap | – | Ball-on-disk nanotribometer | [71] |
| Jatropha | Lithium soap | ZDDP | Four-ball tester | [64] |
| Karanja | Lithium-12-hydroxy stearate | – | Four-ball tester | [54] |
| Rapeseed | Calcium 12-hydroxystearate lithium 12-hydroxystearate | Zinc dioctyldithiophosphate | – | [72] |
| Rapeseed lard | Sodium soap lithium soap | Biological antiwear additive LZ | Four-ball tester | [73] |
| Castor | Chemically modified biopolymers (methylcellulose, chitin, and cellulosic pulp) | | Ball-on-three plates | [68] |
| Sunflower<br>Castor<br>Sunflower+castor<br>glycerol | Beeswax,<br>corncob grits,<br>natural cellulose,<br>lignin | – | Ball-on-disk | [69] |
| Castor | Epoxide-functionalized alkali lignin-based thickener | – | Ball-on-three plates | [70] |

widely used in bearing where greases are subjected to severe conditions. Therefore, the rheological investigation is a significant aspect of the lubrication behavior of the grease. There are many renowned non-Newtonian rheology models viz. the Herschel–Bulkley, power law, Cross model, or Sisko models, which are utilized to assess the rheological characteristics of grease. Enormous literature is available on rheological investigations of conventional greases [59,74–77], but the present chapter focuses on the rheological behavior of vegetable oil-based greases.

García-Zapateiro et al. [65] have formulated green lubricating greases with a combination of base stocks (castor oil, ricinoleic acid-derived estolide, and high-oleic sunflower acid oil (HOS-AO)) and biopolymers (kraft cellulose pulp and chitosan) and concluded that the rheological behavior depends on the viscosity of base stock, polar interactions, and probable chemical reactions amid thickener and base stock. In addition, chitosan-based grease demonstrated higher values of linear viscoelastic functions among all the grease samples. Núñez et al. [78] have also reported that the rheological response of biodegradable grease formulated with castor oil and dispersion of the blend of ethyl cellulose/Kraft cellulose pulp were affected by the concentration and different cellulosic pulp derivative. Sánchez et al. [67] have obtained biodegradable grease using castor oil and cellulose derivatives (viz. ethyl cellulose, methyl cellulose, cellulose acetate α–cellulose). The rheological assessment of castor oil-based greases shows a linear viscoelastic function with frequency and is relatively similar to conventional lithium grease. In addition, the leakage tendency was slightly higher and the mechanical stability is lower as compared to conventional greases. Martín-Alfonso et al. [66] have also pointed out that the consistency and linear viscoelastic functions of biodegradable grease (castor/ethyl cellulose dispersion) are very close to the conventional lithium greases. The above-cited pieces of literature revealed that rheological responses of biodegradable lubricating greases are linear viscoelastic functions with frequency, mechanical stability, and leakage tendency and are found to be approximately similar to those of the conventional lubricating greases. It indicates that biopolymers are a suitable substitute for conventional thickeners.

## 4.11 CONCLUSIONS

This chapter unwraps with a discussion on grease composition then followed by describing the role of additives in the lubrication mechanism. In brief, the synthesis technique of grease is also mentioned. The rapid depletion, unsustainability, and disposal of crude oil products are significant concerns. Biodegradability, renewability, and the eco-friendly nature of vegetable oils are required to consider them as alternative sources to conventional lubricants. Furthermore, this chapter discusses the importance of bio-based greases over conventional greases. The understanding of the tribological and rheological properties of vegetable oil-based greases is vital to know the characteristic properties of these greases. This overview suggests that vegetable oil-based greases have significant potential to protect the tribopairs similar to conventional greases. For sustainable development, it is believed that these vegetable oil-based greases are a substitute for conventional grease.

## 4.12 FUTURE WORK

The present chapter furnishes details about the structure of the triglycerides of vegetable oils and the physicochemical, tribological, and rheological performance of the bio-greases. The results of tribological and rheological investigations of bio-greases showed that vegetable oil-based greases have significant potential to substitute conventional greases. Still, adequate research work is required to enhance the tribological and rheological performance of the bio-greases. It includes:

- Synthesis of biodegradable greases that replace conventional thickener and evaluation for rheological and tribological performance with variable doses of a variety of nanoadditives.
- The study of the effect of various operating parameters on bio-greases with/without nanoadditives.
- The blending of different vegetable oils in the formulation of bio-greases to achieve a satisfactory level of performance.

## REFERENCES


1. Holmberg K, Andersson P, Nylund N, Mäkelä K, Erdemir A. Global energy consumption due to friction in trucks and buses. *Tribiol Int* 2014;78:94–114. doi:10.1016/j.triboint.2014.05.004.
2. Holmberg K, Siilasto R, Laitinen T, Andersson P, Jäsberg A. Global energy consumption due to friction in paper machines. *Tribol Int* 2013;62:58–77. doi:10.1016/j.triboint.2013.02.003.
3. Holmberg K, Andersson P, Erdemir A. Global energy consumption due to friction in passenger cars. *Tribiol Int* 2012;47:221–34. doi:10.1016/j.triboint.2011.11.022.
4. Singh T. Grease Production Survey Report. 22nd Lubricating grease conference. NLGI India-Chapter, 2020.
5. Panchal TM, Patel A, Chauhan DD, Thomas M, Patel J V. A methodological review on bio-lubricants from vegetable oil based resources. *Renew Sustain Energy Rev* 2017;70:65–70. doi:10.1016/j.rser.2016.11.105.
6. Zainal NA, Zulkifli NWM, Gulzar M, Masjuki HH. A review on the chemistry, production, and technological potential of bio-based lubricants. *Renew Sustain Energy Rev* 2018;82:80–102. doi:10.1016/j.rser.2017.09.004.
7. Syahir AZ, Zulkifli NWM, Masjuki HH, Kalam MA, Alabdulkarem A, Gulzar M, et al. A review on bio-based lubricants and their applications. *J Clean Prod* 2017;168:997–1016. doi:10.1016/j.jclepro.2017.09.106.
8. Gerpen J Van. Biodiesel processing and production. *Fuel Process Technol* 2005;86:1097–107. doi:10.1016/j.fuproc.2004.11.005.
9. Mannekote JK, Kailas S V. Studies on boundary lubrication properties of oxidised coconut and soy bean oils. *Lubr Sci* 2009;21:355–65. doi:10.1002/ls.
10. Shashidhara YM, Jayaram SR. Vegetable oils as a potential cutting fluid — An evolution. *Tribiol Int* 2010;43:1073–81. doi:10.1016/j.triboint.2009.12.065.
11. Hsien WLY. Utilization of vegetable oil as bio-lubricant and additive. *Towards Green Lubrication in Machining*. Springer, Singapore; 2015, pp. 7–17. doi:10.1007/978-981-287-266-1.
12. Totten GE. *ASM Handbook*, Volume 18: Friction, Lubrication, and Wear Technology. ASM International; Materials Park, OH, 1992.

13. Torbacke M, Rudolphi ÅK, Kassfeldt E. *Lubricants: Introduction to Properties and Performance.* First Edition. John Wiley & Sons, Ltd; Chichester, 2014.
14. Kashyap A, Harsha AP. Tribological studies on chemically modified rapeseed oil with CuO and $CeO_2$ nanoparticles. *Proc Inst Mech Eng Part J J Eng Tribol* 2016;230:1562–71. doi:10.1177/1350650116641328.
15. Gupta RN, Harsha AP, Singh S. Tribological study on rapeseed oil with nano-additives in close contact sliding situation. *Appl Nanosci* 2018;8:567–80. doi:10.1007/s13204-018-0670-7.
16. Hincapié G, Mondragón F, López D. Conventional and in situ transesterification of castor seed oil for biodiesel production. *Fuel* 2011;90:1618–23. doi:10.1016/j.fuel.2011.01.027.
17. Shomchoam B, Yoosuk B. Eco-friendly lubricant by partial hydrogenation of palm oil over Pd/γ -$Al_2O_3$ catalyst. *Ind Crop Prod* 2014;62:395–9. doi:10.1016/j.indcrop.2014.09.022.
18. Honary LAT, Richter E. *Biobased Lubricants and Greases.* John Wiley & Sons, Ltd; 2011. doi:10.1002/9780470971956.ch7.
19. Rudnick LR. *Synthetics, Mineral Oils, and Bio-Based Lubricants Chemistry and Technology.* CRC Press, Taylor & Francis Group, LLC, Boca Raton, FL; 2006.
20. Mang T, Dresel W, editors. *Lubricants and Lubrication.* Second Edition. WILEY-VCH Verlag GmbH & Co. KGaA; Weinheim, 2007.
21. Lugt PM. *Grease Lubrication in Rolling Bearings.* John Wiley & Sons, Ltd.; 2013. doi:10.1201/b19033-24.
22. Ishchuk YL. *Lubricating Grease Manufacturing Technlogy.* New Age International (P) Limited; New Delhi, 2005.
23. Polishuk AT. U.S. Patent 3,620,975 Mixed complex aluminum soap-clay grease composition, 1971.
24. Lugt PM. A review on grease lubrication in rolling bearings. *Tribol Trans* 2009;52:470–80. doi:10.1080/10402000802687940.
25. Stachowiak GW, Batchelor AW. *Engineering Tribology.* Fourth Edition. Butterworth-Heinemann; Oxford, 2013.
26. Salomonsson L, Stang G, Zhmud B. Oil / thickener interactions and rheology of lubricating greases. *Tribol Trans* 2008;50:302–9. doi:10.1080/10402000701413471.
27. Yamamoto Y, Gondo S. Frictional performance of lithium 12-hydroxystearate greases with different soap fibre structures in sliding contacts. *Lubr Sci* 2002;14:349–62.
28. Lansdown AR. *Tribology in Practice Series: Lubrication and Lubricant Selection-A Practical Guide.* Professional Engineering Publishing Limited London and Bury St Edmunds, UK; 2004.
29. Couronne I, Vergne P. Rheological behavior of greases: Part II - Effect of thermal aging, correlation with physico-chemical changes. *Tribol Trans* 2008;43:788–94. doi:10.1080/10402000008982409.
30. Adhvaryu A, Sung C, Erhan SZ. Fatty acids and antioxidant effects on grease microstructures. *Ind Crops Prod* 2005;21:285–91. doi:10.1016/j.indcrop.2004.03.003.
31. Singh J, Kumar D, Tandon N. Development of nanocomposite grease: Microstructure, flow, and tribological studies. *J Tribol* 2017;139:052001–9. doi:10.1115/1.4035775.
32. Rudnick LR, editor. *Lubricants Additives Chemistry and Applications.* Third Edition. CRC Press, Boca Raton, FL; 2017.
33. Rawat SS, Harsha AP, Agarwal DP, Kumari S, Khatri OP. Pristine and alkylated $MoS_2$ nanosheets for enhancement of tribological performance of paraffin grease under boundary lubrication regime. *J Tribol* 2019;141:072102–12. doi:10.1115/1.4043606.
34. Kamel BM, Mohamed A, El Sherbiny M, Abed KA. Tribological behaviour of calcium grease containing carbon nanotubes additives. *Ind Lubr Tribol* 2016;68:723–8. doi:10.1108/ILT-12-2015-0193.

35. Rawat SS, Harsha AP, Deepak AP. Tribological performance of paraffin grease with silica nanoparticles as an additive. *Appl Nanosci* 2019;9:305–15. doi:10.1007/s13204-018-0911-9.
36. Kobayashi K, Hironaka S, Tanaka A, Umeda K, Iijima S, Yudasaka M, et al. Additive effect of carbon nanohorn on grease lubrication properties. *J Japan Pet Inst* 2005;48:121–6. doi:10.1627/jpi.48.121.
37. Dai W, Kheireddin B, Gao H, Liang H. Roles of nanoparticles in oil lubrication. *Tribol Int* 2016;102:88–98. doi:10.1016/j.triboint.2016.05.020.
38. Gulzar M, Masjuki HH, Kalam MA, Varman M, Zulkifli NWM, Mufti RA, et al. Tribological performance of nanoparticles as lubricating oil additives. *J Nanoparticle Res* 2016;18:223. doi:10.1007/s11051-016-3537-4.
39. Rawat SS, Harsha AP. Current and future trends in grease lubrication. In: Katiyar JK, Bhattacharya S, Patel VK, Kumar V, editors. *Automotive Tribology.* First Edition. Springer Nature, Singapore; 2019, pp. 147–82.
40. Zhao G, Zhao Q, Li W, Wang X, Liu W. Tribological properties of nano-calcium borate as lithium grease additive. *Lubr Sci* 2014;26:43–53. doi:10.1002/ls.
41. Shen T, Wang D, Yun J, Liu Q, Liu X. Tribological properties and tribochemical analysis of nano-cerium oxide and sulfurized isobutene in titanium complex grease. *Tribiol Int* 2016;93:332–46. doi:10.1016/j.triboint.2015.09.028.
42. Ge X, Xia Y, Cao Z. Tribological properties and insulation effect of nanometer $TiO_2$ and nanometer $SiO_2$ as additives in grease. *Tribol Int* 2015;92:454–61. doi:10.1016/j.triboint.2015.07.031.
43. Chang H, Kao M, Luo J, Lan C. Synthesis and effect of nanogrease on tribological properties. *Int J Precis Eng Manuf* 2015;16:1311–6. doi:10.1007/s12541-015-0171-5.
44. Rawat SS, Harsha AP, Das S, Deepak AP. Effect of CuO and ZnO nano-additive on the tribological performance of paraffin oil-based lithium grease. *Tribol Trans* 2019;63:90–100. doi:10.1080/10402004.2019.1664684.
45. Delgado MA, Sánchez MC, Valencia C, Franco JM, Gallegos C. Relationship among microstructure, rheology and processing of a lithium lubricating grease. *Chem Eng Res Des* 2005;83:1085–92. doi:10.1205/cherd.04311.
46. Chen J. Tribological properties of polytetrafluoroethylene, nano-titanium dioxide, and nano-silicon dioxide as additives in mixed oil-based titanium complex grease. *Tribol Lett* 2010;38:217–24. doi:10.1007/s11249-010-9593-5.
47. Ge X, Xia Y, Feng X. Influence of carbon nanotubes on conductive capacity and tribological characteristics of poly(ethylene glycol-ran-propylene glycol) monobutyl ether as base oil of grease. *J Tribol* 2015;138:11801–6. doi:10.1115/1.4031232.
48. Wang L, Zhang M, Wang X, Liu W. The preparation of $CeF_3$ nanocluster capped with oleic acid by extraction method and application to lithium grease. *Mater Res Bull* 2008;43:2220–7. doi:10.1016/j.materresbull.2007.08.024.
49. Sahoo RR, Biswas SK. Effect of layered MoS2 nanoparticles on the frictional behavior and microstructure of lubricating greases. *Tribol Lett* 2014;53:157–71. doi:10.1007/s11249-013-0253-4.
50. ASTM D217-02 Standard Test Methods for Cone Penetration of Lubricating Grease. ASTM International, West Conshohocken, PA; 2003.
51. ASTM D1403-10 Standard Test Methods for Cone Penetration of Lubricating Grease Using One-Quarter and One-Half Scale Cone Equipment. ASTM International, West Conshohocken, PA; 2010. doi:10.1520/D1403-10.2.
52. ASTM D566-16 Standard Test Method for Dropping Point of Lubricating Grease. ASTM International, West Conshohocken, PA; 2016. doi:10.1520/D0566-16.In.
53. ASTM D2265-15 Standard Test Method for Dropping Point of Lubricating Grease Over Wide Temperature Range. ASTM International, West Conshohocken, PA; 2017. doi:10.1520/D2265-15.2.

54. Panchal T, Chauhan D, Thomas M, Patel J. Bio-based grease A value added product from renewable resources. *Ind Crops Prod* 2015;63:48–52. doi:10.1016/j.indcrop.2014.09.030.
55. Florea O, Luca M, Constantinescu A, Florescu D. The influence of lubricating fluid type on the properties of biodegradable greases. *J Synth Lubr* 2003;19:303–13.
56. Adhvaryu A, Erhan SZ, Perez JM. Preparation of soybean oil-based greases: Effect of composition and structure on physical properties. *J Agric Food Chem* 2004;52:6456–9.
57. Sharma BK, Adhvaryu A, Perez JM, Erhan SZ. Soybean oil based greases: Influence of composition on thermo-oxida and thermochemical behaviour. *J Agric Food Chem* 2005;53:2961–8.
58. Sharma BK, Adhvaryu A, Perez JM, Erhan SZ. Biobased grease with improved oxidation performance for industrial application. *J Agric Food Chem* 2006;54:7594–9. doi:10.1021/jf061584c.
59. Rawat SS, Harsha AP, Chouhan A, Khatri OP. Effect of graphene-based nanoadditives on the tribological and rheological performance of paraffin grease. *J Mater Eng Perform* 2020;29:2235–47. doi:10.1007/s11665-020-04789-8.
60. Ji X, Chen Y, Zhao G, Wang X, Liu W. Tribological properties of $CaCO_3$ nanoparticles as an additive in lithium grease. Tribol Lett 2011;41:113–9. doi:10.1007/s11249-010-9688-z.
61. Barriga J, Igartua A, Aranzabe A. Sunflower based grease for heavy duty applications. *Proc World Tribol Congr III - WTC* 2005, 2005, 12–3.
62. Sukirno, RF, Bismo S, Nasikin M. Biogrease based on palm oil and lithium soap thickener: Evaluation of antiwear. *World Appl Sci J* 2009;6:401–7.
63. Sukirno, Ludi, Fajar R, Bismo, Nasikin. Anti-wear properties of bio-grease from modified palm oil and calcium soap thickener. *Agric Eng Int CIGR J* 2010;12:64–9.
64. Nagendramma P, Kumar P. Eco-friendly multipurpose lubricating greases from vegetable residual oils. *Lubricants* 2015;3:628–36. doi:10.3390/lubricants3040628.
65. García-Zapateiro LA, Valencia C, Franco JM. Formulation of lubricating greases from renewable basestocks and thickener agents: A rheological approach. *Ind Crops Prod* 2014;54:115–21. doi:10.1016/j.indcrop.2014.01.020.
66. Martín-Alfonso JE, Núñez N, Valencia C, Franco JM, Díaz MJ. Formulation of new biodegradable lubricating greases using ethylated cellulose pulp as thickener agent. *J Ind Eng Chem* 2011;17:818–23. doi:10.1016/j.jiec.2011.09.003.
67. Sánchez R, Franco JM, Delgado MA, Valencia C, Gallegos C. Development of new green lubricating grease formulations based on cellulosic derivatives and castor oil. *Green Chem* 2009;11:686–93. doi:10.1039/b820547g.
68. Gallego R, Cidade T, Sánchez R, Valencia C, Franco JM. Tribological behaviour of novel chemically modified biopolymer-thickened lubricating greases investigated in a steel-steel rotating ball-on-three plates tribology cell. *Tribol Int* 2016;94:652–60. doi:10.1016/j.triboint.2015.10.028.
69. Acar N, Franco JM, Kuhn E, Gonc DEP, Seabra JHO. Tribological investigation on the friction and wear behaviors of biogenic lubricating greases in steel – Steel contact. *Appl Sci* 2020;10:1–18. doi:10.3390/app10041477.
70. Delgado MA, Cortes-Trivino E, Valencia C, Franco JM. Tribological study of epoxide-functionalized alkali lignin-based gel-like biogreases. *Tribol Int* 2020;146:106231. doi:10.1016/j.triboint.2020.106231.
71. Fiedler M, Kuhn E, Franco JM, Litters T. Tribological properties of greases based on biogenic base oils and traditional thickeners in sapphire-steel contact. *Tribol Lett* 2011;44:293–304. doi:10.1007/s11249-011-9848-9.
72. Buczek B, Zajezierska A. Biodegradable lubricating greases containing used frying oil as additives. *Ind Lubr Tribol* 2015;67:315–9. doi:10.1108/ILT-07-2013-0082.
73. Padgurskas J, Rukuiža R, Kupčinskas A, Kreivaitis R. Lubrication properties of modified lard and rapeseed oil greases with sodium and lithium thickeners. *Ind Lubr Tribol* 2015;67:557–63. doi:10.1108/ILT-12-2014-0140.

74. Kauzlarich JJ, Greenwood JA. Elastohydrodynamic lubrication with Herschel-Bulkley model greases. *ASLE Trans* 1972;15:269–77.
75. Yonggang M, Jie Z. A rheological model for lithium lubricating grease. *Tribol Int* 1998;31:619–25. doi:10.1016/S0301-679X(98)00083-8.
76. Delgado MA, Franco JM. Effect of rheological behaviour of lithium greases on the friction process. *Ind Lubr Tribol* 2008;60:37–45. doi:10.1108/00368790810839927.
77. Franco JM, Delgado MA, Valencia C, Sánchez MC, Gallegos C. Mixing rheometry for studying the manufacture of lubricating greases. *Chem Eng Sci* 2005;60:2409–18. doi:10.1016/j.ces.2004.10.042.
78. Núñez N, Martín-Alfonso JE, Valencia C, Sánchez MC, Franco JM. Rheology of new green lubricating grease formulations containing cellulose pulp and its methylated derivative as thickener agents. *Ind Crops Prod* 2012;37:500–7. doi:10.1016/j.indcrop.2011.07.027.

# 5 Performance Investigation of Hydrodynamic Journal Bearing under Severe Environment

*Sandeep Soni*
S.V. National Institute of Technology

CONTENTS



## 5.1 INTRODUCTION

Larger-sized journal bearings are used to support heavier rotor loads at higher operative speeds in industrial turbo-machinery. These hydrodynamic bearings are engaged in a turbulent regime owing to larger surface speeds. In the open literature, Reynolds numbers vary through 9000 and further have been cited by various researchers for the analysis of turbulence in the hydrodynamic bearings. With the most modern

DOI: 10.1201/9781003096443-5

trends of industrial turbo-machinery, higher Reynolds numbers are possibly featured in the upcoming application of industrial bearings. Accordingly, in the advanced design of journal bearing, turbulence plays a very imperative role in the operative behavior and characteristics of journal bearings.

The high-speed lubricant requirement of industrial machinery has been improved by lubricant additives. Dilatants or pseudoplastic lubricants are used as polymer-thickened oils in journal bearing applications. The additive viscosity in polymer-thickened oil varies and ensures nonlinearity of stress and strain (shear) rates. This correlation can usually be modeled as the 'cubic shear stress model'.

Upcoming designs of hydrodynamic bearing for industrial turbo-machinery of larger sizes and higher speeds are encountered with higher Reynolds numbers and a variety of fluid lubricants.

Constantinescu [1,2] utilized the mixing length hypothesis proposed by Prandtl. He presumes that the mean fluid inertia stresses are imperceptible compared to the fluctuating inertia stresses, called turbulent stresses. The author has developed relations for velocity distribution in journal bearing with the influence of Reynolds number on the turbulent behavior of such bearings. Ng [3] thoroughly explored the approach of turbulent film lubrication affirmed by Constantinescu's approximation of Prandtl's mixing length axiom. The author has also recommended the approximate values of the numerical constants of Reichardt's formula. Ng and Pan [4] revealed the approach of linear turbulence in journal bearing by utilizing Boussinesq's concept of turbulent viscosity in close proximity of wall. They have considered isotropic turbulence and Couette flow to achieve the linearization in the flow through the bearing. Orcutt and Arwas [5] calculated the performance characteristics of a partial arc (100 deg. arc) and a circular bearing in a laminar and turbulent flow regime maintaining the linearized turbulence lubrication approach of Ng and Pan [4]. They have numerically solved the Reynolds lubrication equation to obtain the steady-state and dynamic performance of particular bearings.

Sinhasan and Goyal [6] obtained the transient response of hydrodynamic bearing considering the cubic shear stress model of non-Newtonian lubricants. They have solved 3D Navier–Stokes equations, together with continuity equations, through finite element analysis. Authors have adopted an iteration scheme to precisely simulate the non-Newtonian lubricant performance of hydrodynamic journal bearing. Booser et al. [7] investigated the turbulent behavior (Reynolds number up to 7000) of elliptical and pad-type steam turbine journal bearings up to 5000 rpm. They have established that load acted on the bearing is strongly dependent on the transition speed in between the laminar and turbulent regimes. Hirs [8] expressed the turbulence behavior of a fluid film using the bulk-flow theory. He has employed experimental data to exhibit the relation between shear stress and Reynolds number. Taylor and Dowson [9] documented the popular turbulent lubrication hypotheses of Constantinescu and Ng–Pan–Elrod in a simplified way. They have chalked out their application for fluid-film bearing designers and pointed out that the Ng–Pan–Elrod approach is more accurate than Constantinescu's approach. They have also furnished performance predictions for slider bearing using turbulent lubrication approaches. Soni et al. [10] applied the linear turbulence hypothesis of Ng and Pan to modify the Reynolds lubrication equation using the finite element approach. They have figured out the performance characteristics of a non-circular bearing in the laminar and turbulent regime.

Singh et al. [11] utilized Elrod and Ng's approach of nonlinear turbulence in lubricant films. They have evaluated the performance behavior of a finite circular bearing at Reynolds number up to 13,300 and also compared the performance characteristics with the linear turbulence hypothesis of Ng and Pan. Soni et al. [12] applied Elrod and Ng's nonlinear turbulence theory to the modified Navier–Stokes equations and numerically calculated the performance parameters of a finite bearing for Reynolds numbers up to 13,300. Shenoy and Pai [13] used a linear turbulence model and conception of adjustability in the profile of journal bearing geometry to acquire the performance characteristics in the laminar and turbulent zone.

Nicodemus and Sharma [14] presented analytically the capillary-compensated, four-pocket, hybrid worn journal bearing with various geometric recess shapes in the turbulent regime. The bearing surface wear was modeled through Dufrane's abrasive wear model and turbulence in fluid flow was simulated with Constantinescu's turbulence approach. The author has reported that square recessed shape bearing was better than other recess-shaped bearing from the lubricant film stiffness consideration. Jain and Sharma [15] investigated the effect of journal's geometric imperfections on the steady-state and dynamic characteristics of two-lobed hybrid journal bearing using the power-law model of non-Newtonian lubricants. The authors have reported that the barrel shape irregularity and ellipticity ratio greatly influences the dynamic characteristics of the proposed bearing. Jain and Sharma [16] studied the capillary-compensated, four-lobed multi-recess hybrid bearing in the turbulent flow regime to display the effects of particular geometric irregularities in the journal. The authors have demonstrated the better stability margins obtained in the performance analysis. Rajput and Sharma [17] examined the performance of four-pocket, hybrid journal bearing with a combined effect of misalignment and geometric imperfections of various shapes. These authors have reported that performance characteristics of a proposed bearing degrade due to misalignment and imperfections. Tanner [18] utilized the non-Newtonian power-law model to simulate the lubricant film reactions and friction force for short approximated journal bearing by modifying the Reynolds equation.

Wada and Hayashi [19,20] calculated the steady-state characteristics of finite bearings using the perturbation method by modifying the Reynolds lubrication equation by adopting the cubic shear stress model. Tayal et al. [21] utilized the power-law model and modified 3D-momentum as well as continuity equations to simulate the steady-state performance characteristics of a finite bearing. Tayal et al. [22] simulated the cubic shear stress model of non-Newtonian lubrication to study the performance of a non-circular (elliptical) bearing and figured out the influence of nonlinear factors on the performance behavior of a non-circular bearing. Sinhasan and Goyal [23] presented the transient response of two-lobed hydrodynamic bearing considering the cubic shear stress model of non-Newtonian lubricants. They have solved 3D Navier–Stokes equations, together with continuity equations, through finite element analysis. They have plotted nonlinear journal motion trajectories to precisely simulate the non-Newtonian lubricant performance of the two-lobed hydrodynamic bearing. Hayashi et al. [24] applied polynomial terms for the non-Newtonian aspect of dilatant and pseudoplastic fluids in the modified Reynolds equation to attain circumferential pressure variation and load-carrying capability of hydrodynamic bearings. Safar [25] analytically derived the modified Reynolds lubrication equation by pre-assuming the polynomial expression for velocity distribution. The author obtained the lubricant's pressure variation and load-carrying capacity for various values of eccentricity ratio utilizing the non-Newtonian power-law flow model.

Raghunandana and Majumdar [26] investigated the stability aspect of a circular bearing by utilizing the nonlinear transient method with respect to time. They utilized the Dien and Elrod hypothesis of non-Newtonian film lubricants in their analysis. A higher power-law index numerical value improves the stability of finite bearing. Jang and Chang [27] displayed the use of the power-law model in the adiabatic solution of finite misaligned journal bearing. They assumed oil film viscosity as an exponential function in terms of temperature. They have outlined the effect of the power-law index and misalignment angles on the adiabatic solution of misaligned journal bearing. Jang and Chang [28] analyzed the adiabatic solutions for finite bearing by using the power-law model. These authors have graphically plotted the performance characteristics for power-law index values in the range of 0.7–1.2 and length-to-diameter ratio of 0.5, 1.0, and 1.2 at distinct values of rotational speed. They have also presented the consequences of shear thinning and thickening fluids on the performance of the finite journal bearing. Javorova et al. [29] simulated the Rabinowitsch rheological flow model and elastic deformation of liner in the performance of a finite bearing. They have observed increment in the load-carrying capacity and fluid-film pressure for dilatant lubricants and opposite trends for the pseudoplastic lubricants. Bhujappa and Mareppa [30] analyzed an inclined stepped composite bearing with a Rabinowitsch fluid model to analytically calculate the static and dynamic characteristics of the bearing. The authors have used the perturbation technique to solve the modified Reynolds equation. They have found that non-Newtonian lubricants have significantly affected the performance characteristics. Chetti [31] presented a numerical simulation of journal bearing considering the combined response of elastic deformation and turbulence operated under a couple of stress fluids. The author has outlined that the proposed combination greatly improves the performance of a finite bearing. Hayashi [32] reviewed the various rheological models of non-Newtonian lubricants in the context of hydro-dynamically lubricated bearings. Sheeja and Prabhu [33] theoretically and experimentally executed the performance of plain journal bearing with the impact of thermal effects on non-Newtonian lubrication. They revealed that both non-Newtonian and thermodynamic effects have a strong correlation for the friction in the journal bearing. Das and Guha [34] investigated the performance of micropolar fluid lubricated bearing under the combined influence of turbulence and journal misalignment. Their results suggest that the misalignment moment and turbulence have a greater impact on the load-carrying capacity and friction in the journal bearing. Derdouri and Carreau [35] presented experimentally that at high bearing rotational speeds, viscosity losses as a result of shear thinning cannot coup up with elastic properties of polymer-thickened oils.

The Navier–Stokes and continuity equations under laminar and Newtonian fluids are utilized in the current simulation. The turbulent and non-Newtonian effects are modeled through the hypothesis proposed by Ng and Pan and the cubic shear stress model, respectively. The steady-state performance parameters for a circular bearing with finite approximation in the turbulent regime utilizing a non-Newtonian fluid have been used to compute load capacity, friction coefficient parameter, temperature rise variable, and total oil flow. The obtained results from the present simulation are truly advantageous to the hydrodynamic bearing manufacturers.

## 5.2 CIRCULAR BEARING ANALYSIS

### 5.2.1 Linear Turbulence Modeling

Boussinesq [3] proposed the hypothesis of turbulent fluid motion and according to the proposed theory, the two-dimensional turbulent fluid flow shear stress (effective) can be obtained as [3]:

$$\tau_{ij} = \mu\left(1+\frac{\in}{v}\right)\left(\frac{\partial U_i}{\partial x_j}+\frac{\partial U_j}{\partial x_i}\right) \tag{5.1}$$

where $U$ is the velocity vector. Reichardt suggested the formulation for eddy diffusivity, which could be deduced by [3] using a subsequent relationship [5,11]:

$$\left(\frac{\in}{v}\right) = \dot{k}\left[\Pi^+ - \delta_\ell^+ \tanh\frac{\Pi^+}{\delta_\ell^+}\right] \tag{5.2}$$

where $\delta_\ell^+ = 10.7$, $\dot{k} = 0.4$ are being utilized in the present work, as suggested by [3], and $\Pi^+$ is the dimensionless distance estimated from the nearer wall, which is calculated by a subsequent relationship:

$$\Pi^+ = \begin{Bmatrix} \xi\dfrac{h}{v}\left(\dfrac{|\tau|}{\rho}\right)^{0.5} & 0 < \xi = \dfrac{\Pi^+}{h} \le \dfrac{1}{2} \\ (1-\xi)\dfrac{h}{v}\left(\dfrac{|\tau|}{\rho}\right)^{0.5} & \dfrac{1}{2} < \xi = \dfrac{\Pi^+}{h} \le 1 \end{Bmatrix} \tag{5.3}$$

The Reynolds number Re and local shear stress $|\tau|$ influenced the turbulent fluid behavior near the wall. $|\tau|$ is obtained using the following relationship [36]:

$$|\tau| = \left[(\tau_{\theta r})^2 + (\tau_{zr})^2\right]^{1/2}$$

$$|\tau| = \left[\left\{\mu\left(1+\frac{\in}{v}\right)\left(\frac{\partial u}{\partial r}\right)\right\}^2 + \left\{\mu\left(1+\frac{\in}{v}\right)\left(\frac{\partial w}{\partial r}\right)\right\}^2\right]^{1/2}$$

Thus,

$$|\tau| = \mu\left(1+\frac{\in}{v}\right)\left\{\left(\frac{\partial u}{\partial r}\right)^2 + \left(\frac{\partial w}{\partial r}\right)^2\right\}^{0.5} \tag{5.4}$$

Equations (5.1)–(5.4) have been utilized to revise the governing mathematical relationships for obtaining the performance parameters of a hydrodynamic bearing in the turbulence regime.

### 5.2.2 Modeling of a Non-Newtonian Flow

The cubic shear stress model is inured to simulate the performance of non-Newtonian lubricants in the modeling of a hydrodynamic journal bearing. In dimensionless form, the cubic shear stress model is represented as:

$$\tau + \kappa\tau^3 = \dot{\gamma} \tag{5.5}$$

Equation (5.6) can be adapted to Newtonian ($\kappa = 0$), dilatant ($\kappa < 0$), and pseudoplastic ($\kappa > 0$) lubricants for various journal bearing applications. In Equation (5.6), $\kappa$ denotes the nonlinear factor and $\dot{\gamma}$ is expressed as an invariant of shear strain rate for a three-dimensional fluid flow in the clearance spaces of the geometry and is obtained by the following relationship [6,21]:

$$\dot{\gamma} = \left[2 * \left\{\left(\frac{\partial v}{\partial r}\right)^2 + \left(\frac{1}{r}\frac{\partial u}{\partial \theta} + \frac{v}{r}\right)^2 + \left(\frac{\partial w}{\partial z}\right)^2\right\} + \left(\frac{\partial u}{\partial r} + \frac{1}{r}\frac{\partial v}{\partial \theta} - \frac{u}{r}\right)^2 + \left(\frac{\partial u}{\partial z} + \frac{1}{r}\frac{\partial w}{\partial \theta}\right)^2 + \left(\frac{\partial w}{\partial r} + \frac{\partial v}{\partial z}\right)^2\right]^{0.5} \tag{5.6}$$

The lubricant apparent viscosity is $\mu_a$ and is described as:

$$\mu_a = \left(\tau/\dot{\gamma}\right) \tag{5.7}$$

In the present numerical simulation, flow turbulence and non-Newtonian lubrication have been jointly modeled through the Ng–Pan [4] and cubic shear stress models [6] respectively. The reasons to include these models are as follows:

i. Turbulence may arise in journal bearings due to higher operational speed in the modern-day industrial machinery and so far as per the literature review, the turbulence hypothesis of Ng–Pan [4] is more promising in close proximity of severe flow conditions than Constantinescu's turbulence model [1,2].
ii. The experimental investigation of the cubic shear stress model for the hydrodynamic journal bearing has been verified by Wada and Hayashi [20] through illustrating the fluid-film pressure distribution, where the load-carrying capacity is modest compared to the Newtonian lubricants.

## 5.3 FINITE ELEMENT PROCEDURE

The momentum equations (or N-S equation), which administer the Newtonian and incompressible fluid flow in the wedge gap of a finite circular bearing, are revised in the form of turbulence as well as non-Newtonian fluid flow and can be formulated as:

$$\frac{R_1}{r}\frac{\partial p}{\partial \theta}=\mu_a(1+K)\left\{\frac{\partial^2 u}{\partial r^2}+\frac{1}{r}\frac{\partial u}{\partial r}+\frac{1}{r^2}\frac{\partial^2 u}{\partial \theta^2}+\frac{1}{R_1^2}\frac{\partial^2 u}{\partial z^2}+\frac{2}{r^2}\frac{\partial v}{\partial \theta}-\frac{u}{r^2}\right\}$$

$$R_1\frac{\partial p}{\partial r}=\mu_a(1+K)\left\{\frac{\partial^2 v}{\partial r^2}+\frac{1}{r}\frac{\partial v}{\partial r}+\frac{1}{r^2}\frac{\partial^2 v}{\partial \theta^2}+\frac{1}{R_1^2}\frac{\partial^2 v}{\partial z^2}-\frac{2}{r^2}\frac{\partial u}{\partial \theta}-\frac{v}{r^2}\right\} \quad (5.8a)$$

$$\frac{\partial p}{\partial z}=\mu_a(1+K)\left\{\frac{\partial^2 w}{\partial r^2}+\frac{1}{r}\frac{\partial w}{\partial r}+\frac{1}{r^2}\frac{\partial^2 w}{\partial \theta^2}+\frac{1}{R_1^2}\frac{\partial^2 w}{\partial z^2}\right\}$$

The dimensionless form of continuity equation is expressed as:

$$\left\{\frac{1}{r}\frac{\partial u}{\partial \theta}+\frac{v}{r}+\frac{\partial v}{\partial r}+(R_1)^{-1}\frac{\partial w}{\partial z}\right\}=0 \quad (5.8b)$$

where $\mu_a$ is the apparent viscosity, $K=(\in/v)$ denotes the turbulent viscosity coefficient, and $R_1=(R/C)$. The velocities $u$, $v$, and $w$ are in the $\theta$, $r$, and $z$ directions.

The velocities ($u$, $v$, and $w$) and pressure variation ($p$) are obtained from the iterative solution of governing Equations (5.8a and 5.8b) after applying the Reynolds boundary condition. During the present simulation, the convective and local inertia components of films are ignored. Galerkin's technique is applied to Equations (5.8a and 5.8b) in the finite element solution stage and the following equations are obtained for an element:

$$\iiint N_i^u\left\{\frac{R_1}{r}\frac{\partial p}{\partial \theta}-\mu_a(1+K)\left(\frac{\partial^2 u}{\partial r^2}+\frac{1}{r}\frac{\partial u}{\partial r}+\frac{1}{r^2}\frac{\partial^2 u}{\partial \theta^2}+\frac{1}{R_1^2}\frac{\partial^2 u}{\partial z^2}+\frac{2}{r^2}\frac{\partial v}{\partial \theta}-\frac{u}{r^2}\right)\right\}r\,d\theta\,dr\,dz=0$$

$$\iiint N_i^u\left\{R_1\frac{\partial p}{\partial r}-\mu_a(1+K)\left(\frac{\partial^2 v}{\partial r^2}+\frac{1}{r}\frac{\partial v}{\partial r}+\frac{1}{r^2}\frac{\partial^2 v}{\partial \theta^2}+\frac{1}{R_1^2}\frac{\partial^2 v}{\partial z^2}-\frac{2}{r^2}\frac{\partial u}{\partial \theta}-\frac{v}{r^2}\right)\right\}r\,d\theta\,dr\,dz=0 \quad (5.9)$$

$$\iiint N_i^u\left\{\frac{\partial p}{\partial z}-\mu_a(1+K)\left(\frac{\partial^2 w}{\partial r^2}+\frac{1}{r}\frac{\partial w}{\partial r}+\frac{1}{r^2}\frac{\partial^2 w}{\partial \theta^2}+\frac{1}{R_1^2}\frac{\partial^2 w}{\partial z^2}\right)\right\}r\,d\theta\,dr\,dz=0$$

$$\iiint N_j^p\left\{\frac{1}{r}\frac{\partial u}{\partial \theta}+\frac{v}{r}+\frac{\partial v}{\partial r}+(R_1)^{-1}\frac{\partial w}{\partial z}\right\}r d\theta dr dz=0$$

where

$$u^e=\sum_{i=1}^{20}N_i^u u_{i'} \qquad v^e=\sum_{i=1}^{20}N_i^u v_{i'}$$
$$w^e=\sum_{i=1}^{20}N_i^u w_{i'} \qquad p^e=\sum_{j=1}^{8}N_j^p p_{j'} \quad (5.10)$$

The parabolic and linear shape functions (3D isoparametric elements) have been utilized in the discretization for velocity and pressure respectively [21,37]. Appropriate substitutions and integration of Equation (5.10) give an element equation in the matrix form as:

$$\begin{pmatrix} K^u & K'^p \\ K^p & 0 \end{pmatrix}^e \begin{Bmatrix} \Phi_u \\ \Phi_p \end{Bmatrix}^e = 0 \tag{5.11}$$

where $\Phi_u, \Phi_p, K'^p, K^p$, and $K^u$ represent sub-matrix assembly for nodal velocity, nodal pressure, continuity equation, pressure, and viscous terms respectively. The assembly of fluidity matrices for every element provides a global equation in the following form:

$$[K]\{\Phi\} = b \tag{5.12}$$

Hence, the required nodal velocities and nodal pressure are obtained with the iterative solution.

## 5.4 SOLUTION SCHEME

Firstly, nodal pressure and nodal velocities have been computed under laminar flow, and then these values have been upgraded by considering the turbulent viscosity coefficient $(K = (\in /v))$, local shear stress $(|\tau|)$, and different values of Reynolds number $(\text{Re})$. Simultaneously, the solution of Newtonian lubricant has been upgraded with apparent viscosity $(\mu_a)$ and shear strain rate $(\dot{\gamma})$ to consider the effect of non-Newtonian lubricants in the calculation of various performance parameters of finite journal bearing.

In the final phase of discretization, a total of 40 elements (i.e. 4 elements in axial, 10 elements in circumferential, and 1 element in radial directions) and 353 nodes are considered to achieve the accuracy of solution for the steady-state parameters of the finite bearing system under the severe proposed operating conditions.

In such a way, the joint consequence of the turbulence regime and non-Newtonian lubricant has been considered.

Figure 5.1 ('**Block StTurb_NoN**' and '**Block NgPan_CSSM**') depicts the solution scheme for the current numerical simulation.

## 5.5 STEADY-STATE PERFORMANCE PARAMETERS

The steady-state performance parameters of circular journal bearing with aspect ratio 1, in expressions of load capacity, friction coefficient parameter, temperature rise variable, and total oil flow as it may be attained by subsequent relationships.

### 5.5.1 Load Capacity and Attitude Angle

The load capacity of the bearing system is enumerated by combining the fluid-film force components of every element perpendicular and along to the journal bearing centers:

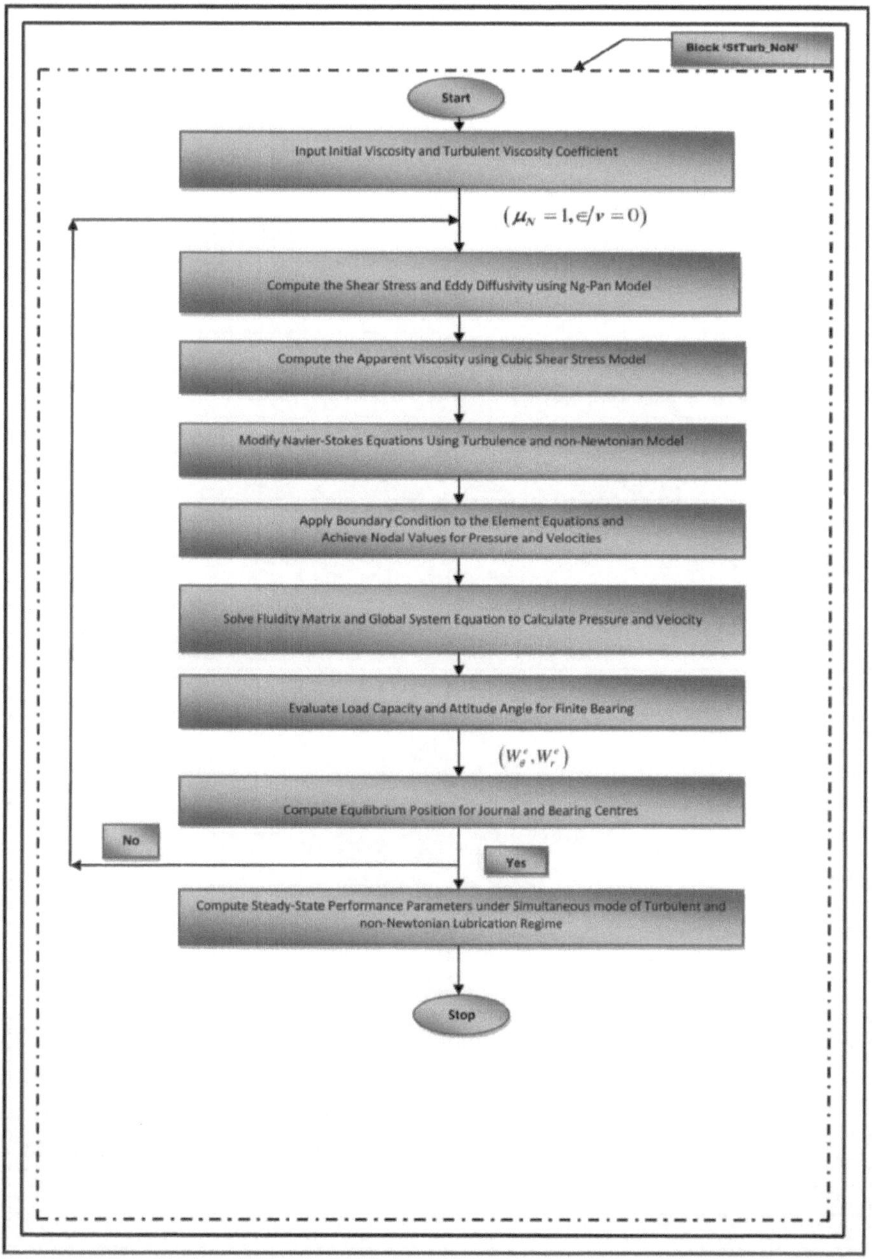


**FIGURE 5.1** (a) Flow diagram for the solution scheme 'Block StTurb_NoN'. (b) Flow diagram for turbulence and non-Newtonian model 'Block NgPan_CSSM'.

*(Continued)*

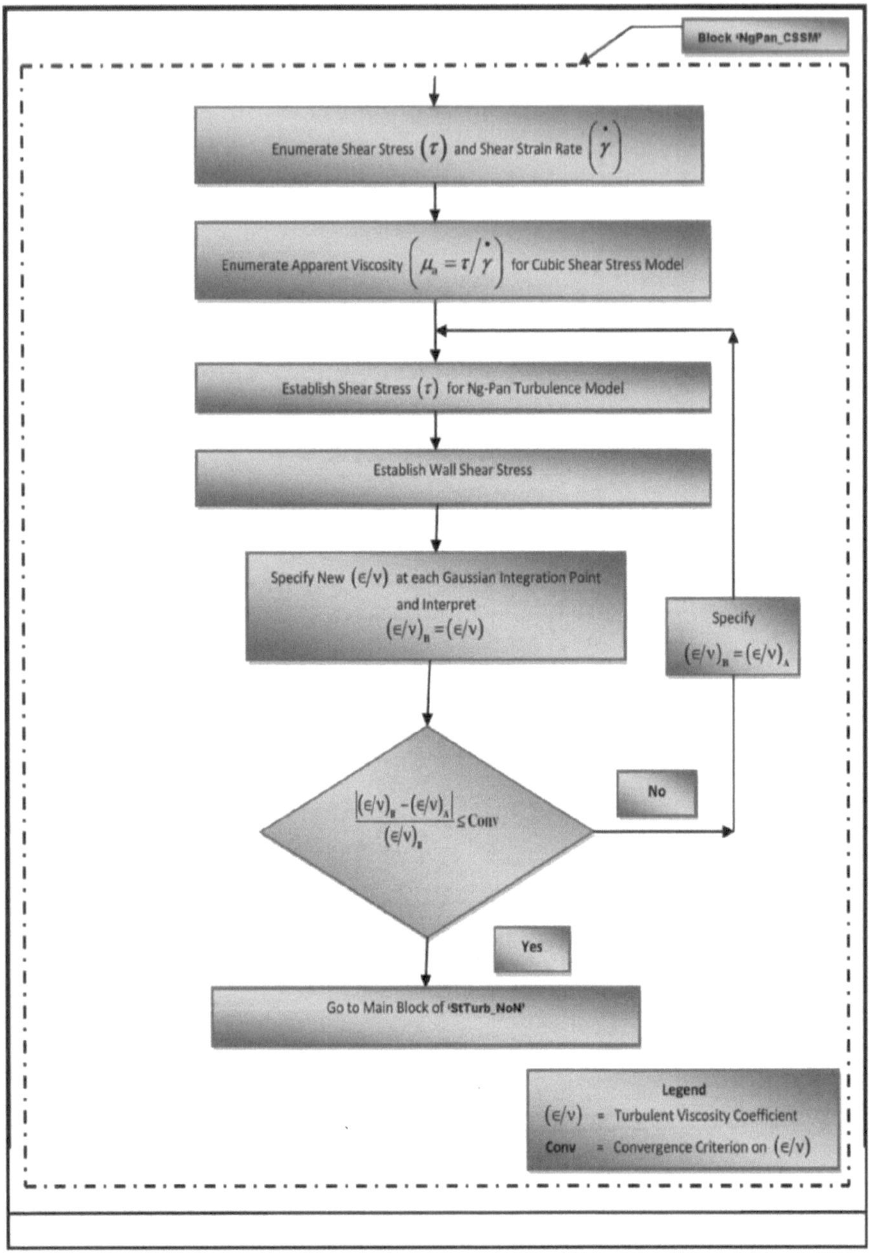


**FIGURE 5.1 (*Continued*)** (a) Flow diagram for the solution scheme 'Block StTurb_NoN'. (b) Flow diagram for turbulence and non-Newtonian model 'Block NgPan_CSSM'.

$$W_\theta = \sum W_\theta^e, \quad W_r = \sum W_r^e,$$

where

$$W_\theta^e = \iint (p)^e \sin\theta \, r \, d\theta \, dz$$

$$W_r^e = -\iint (p)^e \cos\theta \, r \, d\theta \, dz$$

The following relationship provides resultant load capacity, the Sommerfield number, and attitude angle for the proposed bearing:

$$W = \left(W_r^2 + W_\theta^2\right)^{1/2}$$

$$S = \left(\frac{2L}{\pi DW}\right) \tag{5.13}$$

$$\varphi = \tan^{-1}\left(\frac{W_\theta}{W_r}\right)$$

### 5.5.2 Friction Coefficient Parameter

The friction force acted upon the moving surface of journal is considered by summation of friction force on every element, which explicates as:

$$F_j = \sum F_j^e$$

where

$$F_j^e = \iint \left\{ \mu_a \left(1 + \frac{\in}{v}\right) \frac{1}{H} + \frac{H}{2} \frac{\partial p}{\partial \theta} \right\}^e r \, d\theta \, dz$$

The friction coefficient parameter is obtained by the following expression:

$$f\left(\frac{R}{C}\right) = \left(\frac{F_j}{W}\right) \tag{5.14}$$

### 5.5.3 Temperature Rise Variable

The temperature rise variable for the present simulation is calculated as follows:

$$\delta T = \left(\frac{F_j}{Q_z}\right) \tag{5.15}$$

### 5.5.4 Total Oil Flow

The lubricant flow from the journal bearing is obtained by the following expression:

$$Q = \sum q^e \quad q^e = \iint (w)^e \, r d\theta dr \tag{5.16}$$

## 5.6 RESULT AND DISCUSSIONS

A computer algorithm in MATLAB® has been prepared to accurately simulate the steady-state performance parameters of finite circular journal bearing under the joint influence of linear turbulence and non-Newtonian lubrication. In the simulated results, the Reynolds number ranges from the laminar to turbulent regime (Re = 3326, 8314 and 13,300), and the nonlinearity of the cubic shear stress model varies from 0.10 to 1.00. The value of $\kappa = 0$ stands for a Newtonian lubricant.

The validity of computed results verified from references [4] and [6] and the computed outcomes agrees well with selected references for the present simulation.

Figure 5.2a and b depicts the variation of dimensionless load capacity with eccentricity ratio attained from references [4,6] and current simulation (i.e. combined Re, $\kappa$). In general, the dimensionless load capacity increases with an increase in eccentricity ratio, and the combined load capacity with Re = 3326, $\kappa = 0.1$ to 1.0 comes in between linear turbulence and a non-Newtonian lubricant (Figure 5.2a). From Figure 5.2b, it is observed that the combined load capacity with Re = 3326, 8314 and 13,300, $\kappa = 0.1$ increases significantly as compared to linear turbulence and non-Newtonian load capacity. The reason for this variation is that, when the nonlinearity increases in flow, the apparent viscosity decreases, and when the Reynolds number increases, the turbulent viscosity coefficient changes sharply.

Figure 5.3a and b illustrates the change in $f(R/C)$ with $\varepsilon$ attained by the linear turbulence theory [4], non-Newtonian lubrication [6], and current simulation (i.e. combined Re, $\kappa$). The friction coefficient parameter reduces by way of increase in the bearing eccentricity ratio. As nonlinearity increases with constant Reynolds number (Figure 5.3a), the values of $f(R/C)$ are higher than those of separate turbulence and non-Newtonian lubrication; this type of trend is due to decrement in the apparent viscosity of non-Newtonian fluids. By keeping the constant value of lubricant nonlinearity and increment in the flow turbulence (Re = 3326–13,300), the friction coefficient parameter decreases sharply as presented in Figure 5.3b.

Figure 5.4a and b shows the variation of temperature rise variable ($\delta T$) with eccentricity ratio ($\varepsilon$) obtained from references [4,6] and current simulation (i.e. combined Re, $\kappa$). Generally, $\delta T$ decreases with an increase in the eccentricity ratio under the joint effect of linear turbulence and non-Newtonian lubrication. As the temperature rise variable depends upon the frictional force and lubricant flow, it decreases but is little influenced by keeping the constant value of Re and increasing the nonlinearity factor from 0.1 to 1.0 jointly as shown in Figure 5.4a. As the turbulence increases with the constant nonlinearity of non-Newtonian flow, the $\delta T$ values are showing higher trends than the laminar and non-Newtonian flow. Flow transition from laminar to turbulent is responsible for this trend (Figure 5.4b).

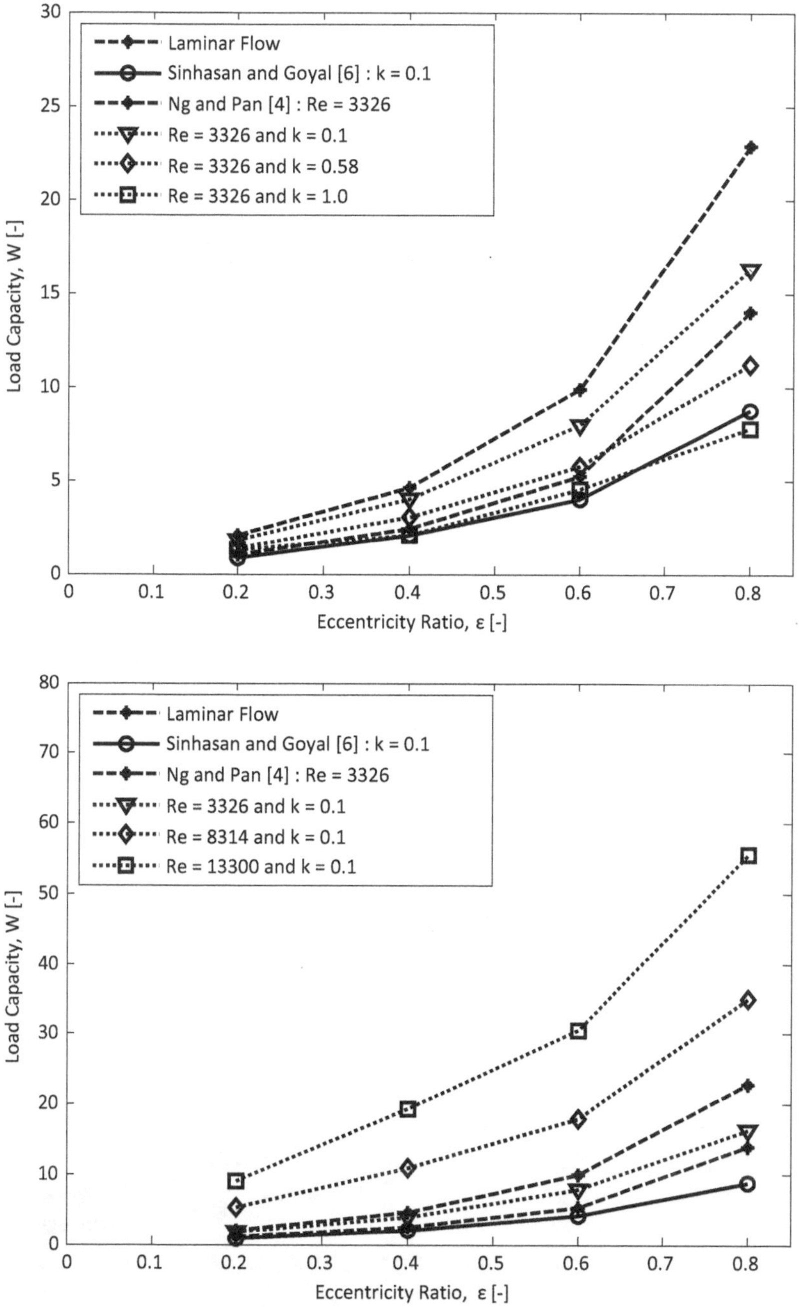


**FIGURE 5.2** (a) Variation of $W$ with $\varepsilon$ (laminar, Re = 3326 and different values of $\kappa$). (b) Variation of $W$ with $\varepsilon$ (laminar, $\kappa = 0.1$ and different values of Re).

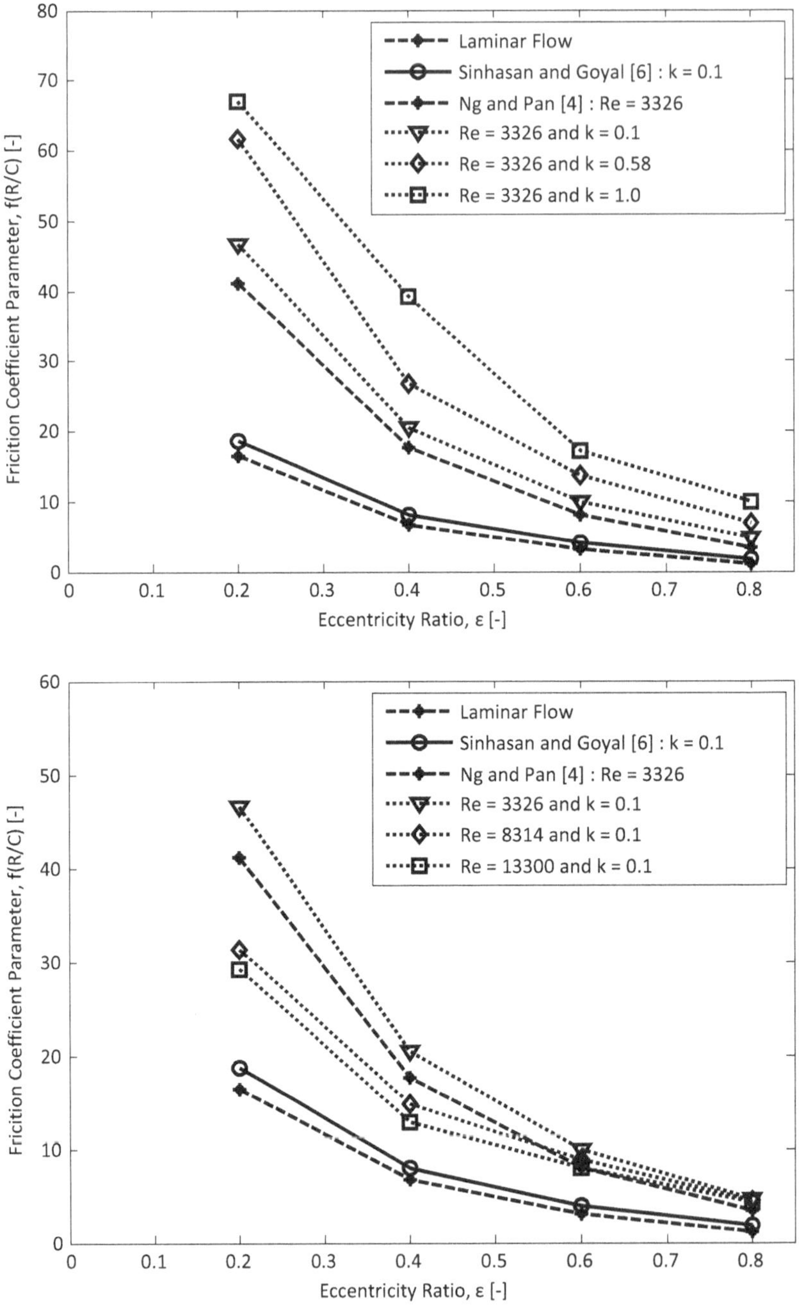


**FIGURE 5.3** (a) Variation of $f(R/C)$ with $\varepsilon$ (laminar, Re = 3326 and different values of $\kappa$). (b) Variation of $f(R/C)$ with $\varepsilon$ (laminar, $\kappa = 0.1$ and different values of Re).

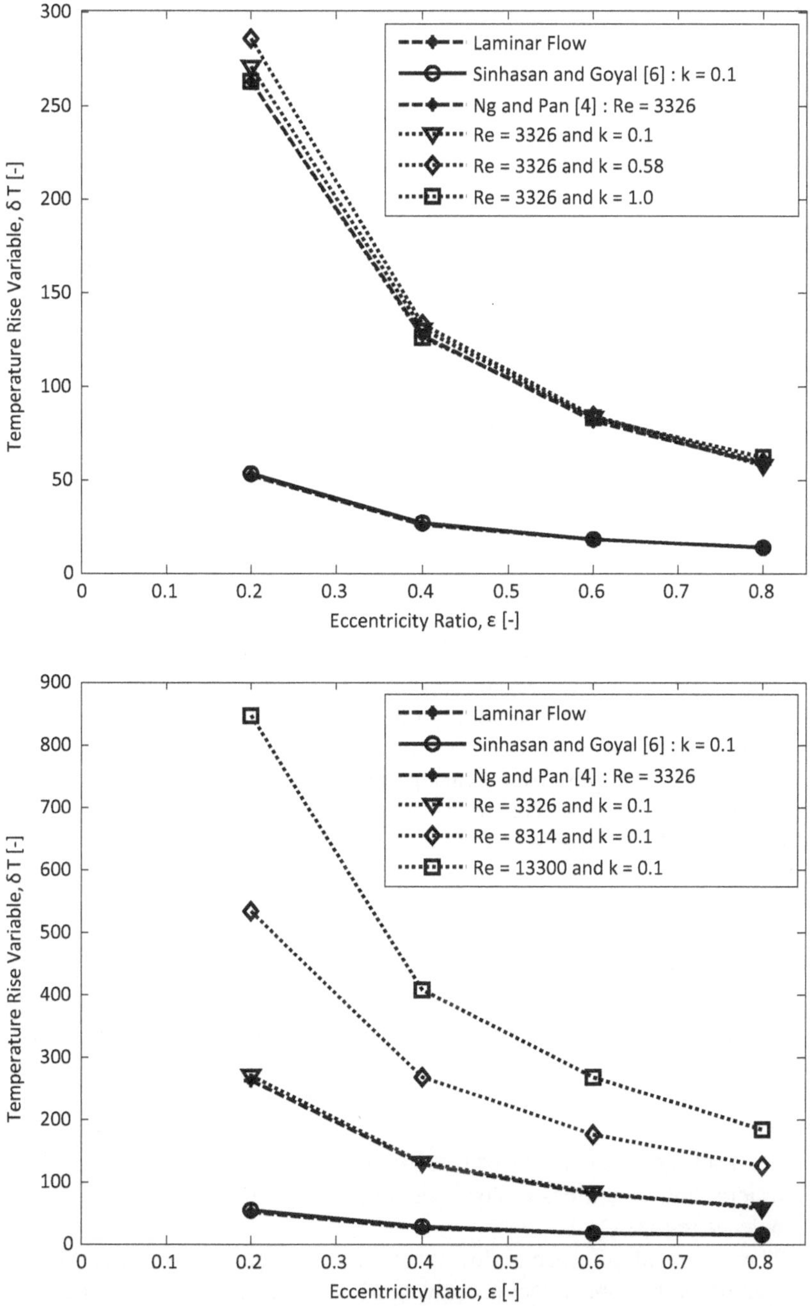


**FIGURE 5.4** (a) Variation of $\delta T$ with $\varepsilon$ (laminar, Re = 3326 and different values of $\kappa$). (b) Variation of $\delta T$ with $\varepsilon$ (laminar, $\kappa = 0.1$ and different values of Re).

At a higher eccentricity ratio, a greater quantity of oil flow is required for the successful operation of the bearing. Under joint flow operating conditions, oil flow ($Q$) is showing a little increasing trend, so the temperature rise variable ($\delta T$) discerns decreasing variation.

Figure 5.5a and b shows the change of total oil flow with eccentricity ratio obtained by linear turbulence [4], non-Newtonian lubrication [6], and present simulation. The total oil flow commonly depicts an increasing trend with the variation of eccentricity ratio as well as with the Reynolds number and nonlinear factor. The joint effect (i.e. Re,$\kappa$) on bearing oil flow is infinitesimal as observed from Figure 5.5a and b.

Figures 5.6–5.9 show the impact of fluid nonlinearity ($\kappa$) at constant Reynolds number (Re = 8314) with various values of $\varepsilon$ from 0.2 to 0.8 on the selected steady-state performance parameters. The rheological behavior of a non-Newtonian fluid is governed by the cubic shear stress model.

Figure 5.6 depicts that initially the load capacity increases and then steadily decreases in the Newtonian and non-Newtonian range respectively. This is due to the fact that the operating fluid viscosity reduces on the increment of nonlinearity so the load capacity is also reduced.

Figure 5.7 shows that the increment in the nonlinearity of lubricant increases the friction coefficient parameter at a particular value of Reynolds number. This variation in $f(R/C)$ at $\varepsilon = 0.8$ with specific Re is minimum.

It can be noted from Figure 5.8 that the $\delta T$ value initially increases in the Newtonian regime and then varies constantly in the non-Newtonian regime. This trend is minimal at a higher eccentricity ratio for Reynolds number 8314.

For a specific value of Re and eccentricity ratio with variation in nonlinearity influences the total oil flow from bearing appreciably. This variation in Figure 5.9 is a result of an increase in eccentricity ratio that generates more pressure gradient axially and thus the oil flow increases.

## 5.7 CONCLUSIONS

The hypothesis of linear turbulence recommended by Ng and Pan [4] in conjunction with the cubic shear stress model of fluid lubrication has been evinced, with FEM to simulate and compute the steady-state performance parameters of the finite circular bearing under a severe working environment of fluid flow. From the results presented in this chapter, the following conclusions can be outlined:

1. The load capacity of a journal bearing system manifests an increasing trend in joint flow conditions as confronted by the Newtonian lubricant and turbulence regime.
2. The combined influence of flow turbulence and non-Newtonian lubrication signifies a favorable effect in figuring out the frictional behavior of a finite bearing.
3. The temperature rise variable in the simultaneous mode of turbulence and non-Newtonian lubricant significantly affects the performance of hydrodynamic bearing.

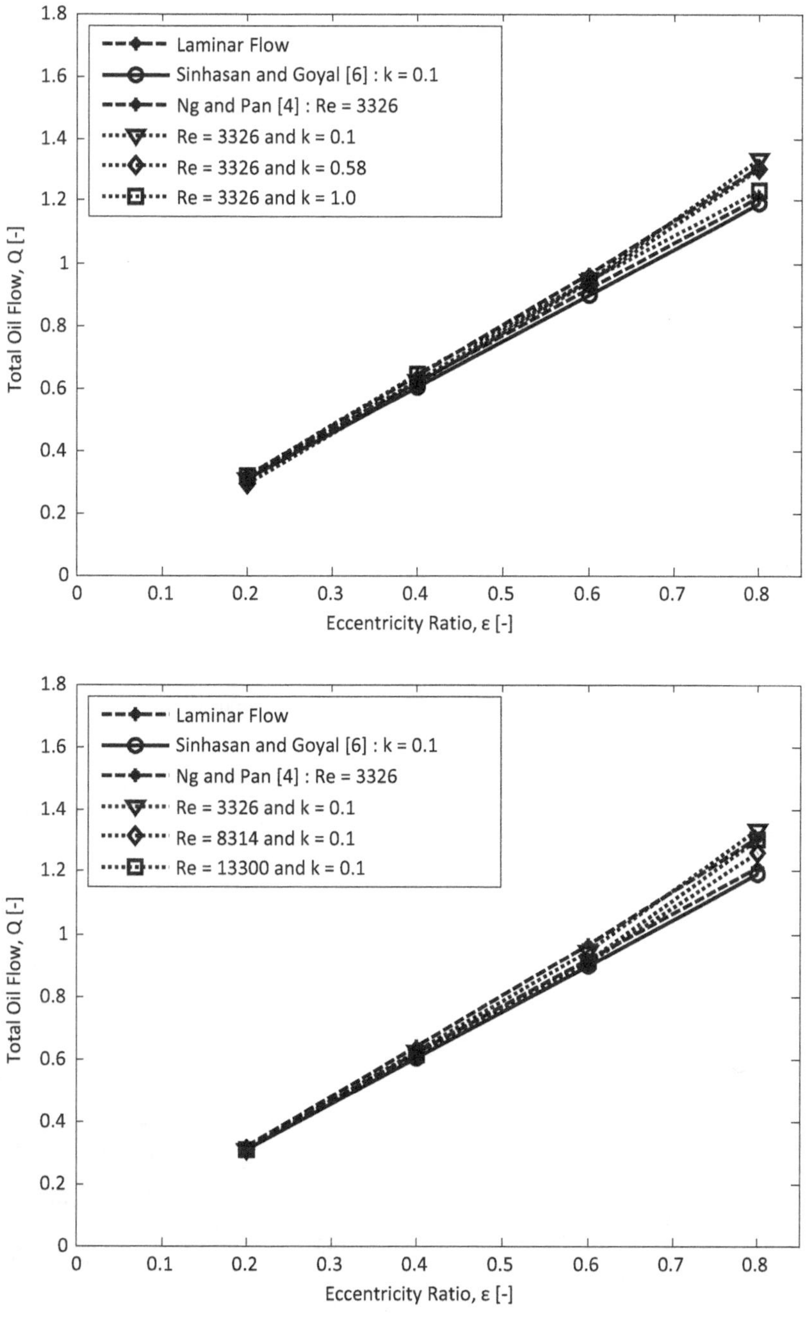


**FIGURE 5.5** (a) Variation of $Q$ with $\varepsilon$ (laminar, Re = 3326 and different values of $\kappa$). (b) Variation of $Q$ with $\varepsilon$ (laminar, $\kappa = 0.1$ and different values of Re).

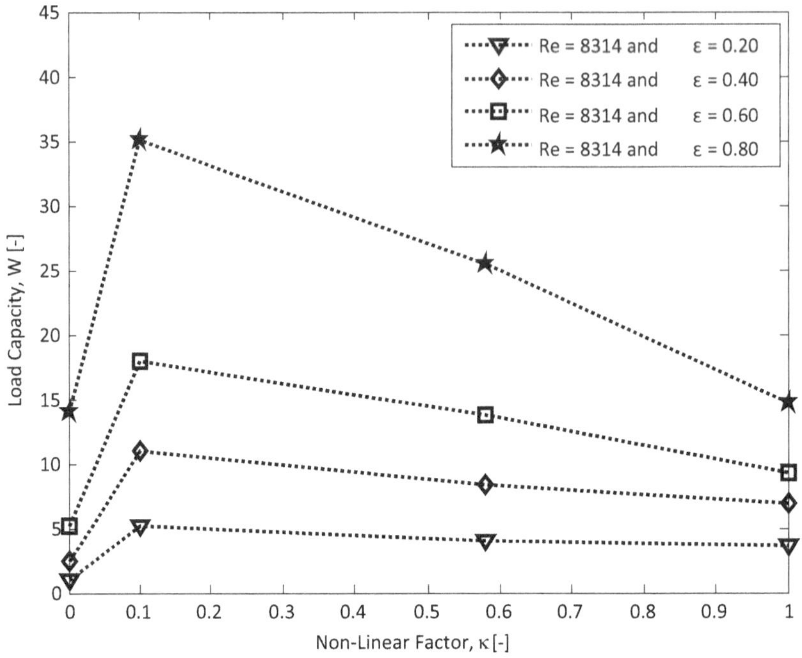


**FIGURE 5.6** Variation of $W$ with $\kappa$ (Re = 8314 and different values of $\varepsilon$).

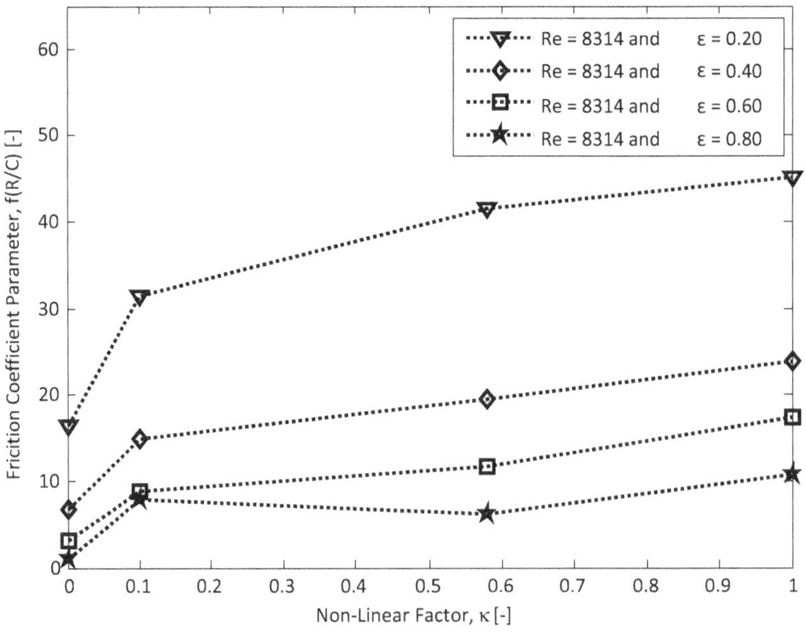


**FIGURE 5.7** Variation of $f(R/C)$ with $\kappa$ (Re = 8314 and different values of $\varepsilon$).

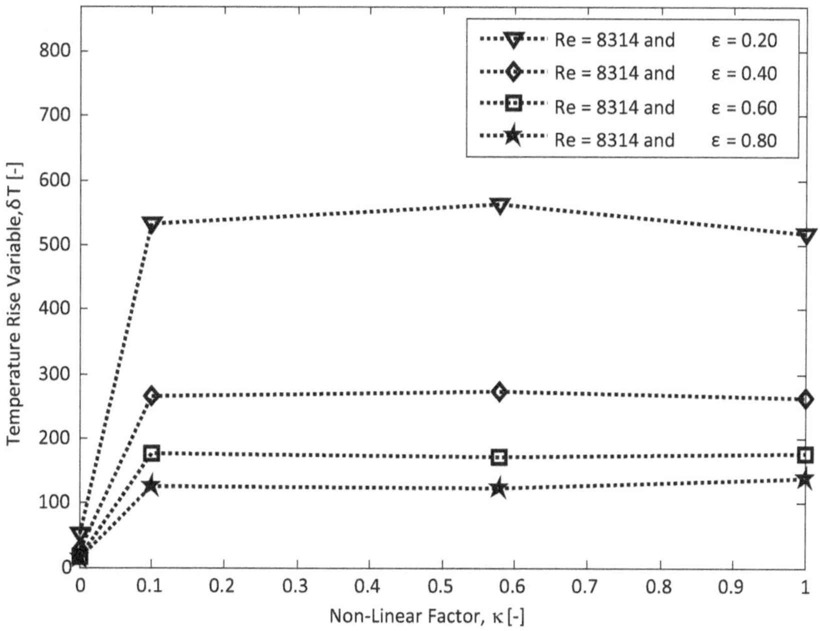


**FIGURE 5.8** Variation of $\delta T$ with $\kappa$ (Re = 8314 and different values of $\varepsilon$).

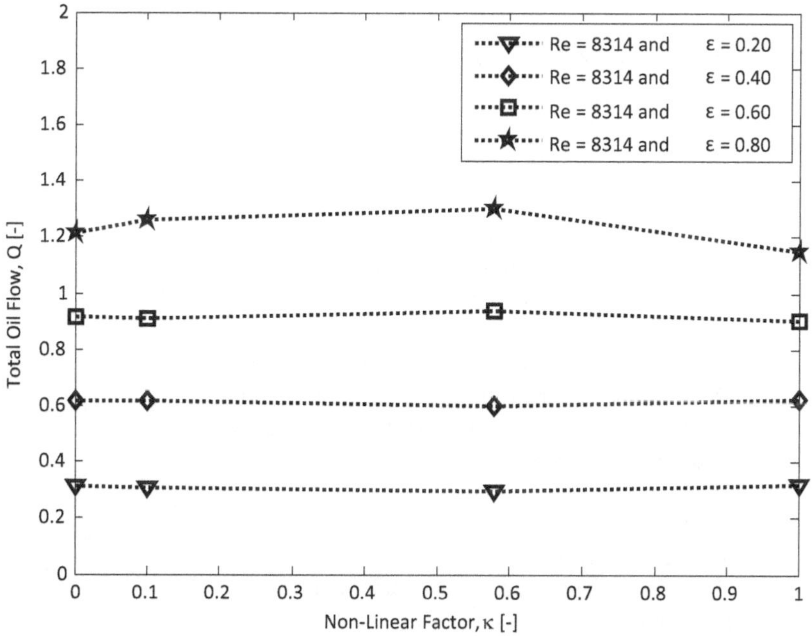


**FIGURE 5.9** Variation of $Q$ with $\kappa$ (Re = 8314 and different values of $\varepsilon$).

4. Total oil flow through the bearing is slightly affected under the proposed working flow environment.
5. The variation in lubricant's nonlinearity with Reynolds number influences the rheological characteristics of finite bearing thoroughly.
6. The concurrent effect of a severe flow environment plays a decisive role in the selection of hydrodynamic journal bearing for industrial turbo-machinery applications.

## NOMENCLATURE

| | |
|---|---|
| $C$ | radial clearance |
| $e$ | bearing eccentricity |
| $R$ | radius of journal |
| $L$ | length of bearing |
| $D$ | diameter of journal |
| $(\bar{u},\bar{v},\bar{w})$ | tangential, radial, and axial velocity |
| $\bar{p}$ | fluid-film pressure |
| $\mu$ | absolute viscosity of lubricant |
| $\nu$ | kinematic viscosity of lubricant |
| $\rho$ | density of lubricant |
| $h$ | oil film thickness |
| $\omega$ | spin velocity of journal |
| $\bar{r}$ | radial coordinate measured from journal center |
| $\bar{z}$ | axial coordinate measured from bearing center |

## DIMENSIONLESS PARAMETERS

| | |
|---|---|
| $p$ | fluid-film pressure |
| $\tau_{ij}$ | effective shear stress |
| $\dot{k}$ | constant = 0.4 |
| $\delta_\ell^+$ | constant = 10.7 |
| $\Pi^+$ | non-dimensional distance measured from the nearest wall |
| $(\in/\nu)$ | eddy diffusivity or turbulent viscosity coefficient |
| $\xi$ | radial coordinate measured from the surface of journal |
| $\varepsilon$ | eccentricity ratio $= e/C$ |
| $\theta$ | circumferential coordinate measured from journal center |
| $r$ | radial coordinate measured from journal center $= \bar{r}/C$ |
| $z$ | axial coordinate measured from bearing center $= \bar{z}/C$ |
| $(u,v,w)$ | $\bar{u}/\omega R, \bar{v}/\omega R, \bar{w}/\omega R$ |
| $\kappa$ | non- linear factor |
| $\dot{\gamma}$ | shear strain rate |
| $\mu_a$ | apparent viscosity |
| $\tau$ | shear stress defined in non-Newtonian law |
| Re | Reynolds number $= (\rho\omega RC/\mu)$ |
| $R_1$ | clearance ratio $= R/C$ |

| | |
|---|---|
| $N_i^u$ | parabolic shape functions for velocity at node $i$ |
| $N_j^p$ | linear shape functions for pressure at node $j$ |
| $W_r$ | oil film force along the line of centers |
| $W_\theta$ | oil film force perpendicular to the line of centers |
| $W$ | resultant load capacity |
| $S$ | Sommerfield number |
| $\varphi$ | attitude angle (deg.) |
| $f(R/C)$ | friction coefficient parameter |
| $Q$ | total oil flow |
| $F_j$ | frictional force |
| $\delta T$ | temperature rise variable |

## MATRICES

| | |
|---|---|
| $\left[K^u\right]$ | sub-matrix for viscous terms for element equation |
| $\left[K^p\right]$ | sub-matrix for continuity equation terms for element equation |
| $\left[K'^p\right]$ | sub-matrix for pressure terms for element equation |
| $[\Phi_u]$ | sub-matrix for velocity variable for element equation |
| $\left[\Phi_p\right]$ | sub-matrix for pressure variable for element equation |
| $[K]$ | fluidity matrix for the entire assembly |
| $\{\Phi\}$ | column matrix for nodal velocities and nodal pressure for the entire assembly |
| $b$ | column matrix for right-hand side variables for the entire assembly |

## REFERENCES


1. Constantinescu, V. N. (1959), "On Turbulent Lubrication," *Proceedings of the Institution of Mechanical Engineers*, **173** (1), pp 881–900.
2. Constantinescu, V. N. (1962), "Analysis of Bearings Operating in Turbulent Regime," *Journal of Fluids Engineering*, **84** (1), pp. 139–151.
3. Ng, C. W. (1964), "Fluid Dynamic Foundation of Turbulent Lubrication Theory," *ASLE Transactions*, **7** (4), pp. 311–321.
4. Ng, C. W. and Pan, C. H. T. (1965), "A Linearized Turbulent Lubrication Theory," *Journal of Basic Engineering*, **87** (3), pp. 675–682.
5. Orcutt, F. K. and Arwas, E. B. (1967), "The Steady-State and Dynamic Characteristics of a Full Circular Bearing and a Partial Arc Bearing in the Laminar and Turbulent Flow Regimes," *Journal of Tribology*, **89** (2), pp. 143–153.
6. Sinhasan, R. and Goyal, K. C. (1992), "Transient Response of A Circular Journal Bearing Lubricated with non-Newtonian Lubricants," *Wear*, **156**, pp. 385–399.
7. Booser, E. R., Missana, A. and Ryan, F. D. (1970), "Performance of Large Steam Turbine Journal Bearings," *ASLE Transactions*, **13** (4), pp. 262–268.
8. Hirs, G. G. (1973), "A Bulk Flow Theory for Turbulence in Lubricant Films," *Journal of Tribology*, **95** (2), pp. 137–145.
9. Taylor, C. M. and Dowson, D. (1974), "Turbulent Lubrication Theory–Application to Design," *Journal of Tribology*, **96** (1), pp. 36–46.

10. Soni, S. C., Sinhasan, R. and Singh, D. V. (1981), "Performance Characteristics of Noncircular Bearings in Laminar and Turbulent Flow Regimes," *ASLE Transactions*, **24** (1), pp. 29–41.
11. Singh, D. V., Sinhasan, R. and Soni, S. C. (1983), "Static and Dynamic Analysis of Hydrodynamic Bearings in Laminar and Superlaminar Flow Regimes by Finite Element Method," *ASLE Transactions*, **26** (2), pp. 255–263.
12. Soni, S. C., Sinhasan, R. and Singh, D. V. (1983), "Analysis by the Finite Element Method of Hydrodynamic Bearings Operating in the Laminar and Superlaminar Regimes," *Wear*, **84**, pp. 285–296.
13. Shenoy, B. S. and Pai, R. (2010), "Stability Characteristics of an Externally Adjustable Fluid Film Bearing in the Laminar and Turbulent Regimes," *Tribology International*, **43**, pp. 1751–1759.
14. Nicodemus, E. R. and Sharma, S. C. (2010), "A Study of Worn Hybrid Journal Bearing System With Different Recess Shapes Under Turbulent Regime," *Journal of Tribology*, **132**, pp. 041704-1–041704-12.
15. Jain, D. and Sharma, S. C. (2015), "Two-lobe Geometrically Imperfect Hybrid Journal Bearing Operating with Power Law Lubricant," *Proc IMechE Part J: Journal of Engineering Tribology*, **229**(1), pp. 30–46.
16. Jain, D. and Sharma, S. C. (2015), "Combined Influence of Geometric Irregularities of Journal and Turbulence on the Performance of Four-Lobe Hybrid Journal Bearing," *Proc IMechE Part J: Journal of Engineering Tribology*, 2015, **229** (12), pp. 1409–1424.
17. Rajput, A. K. and Sharma, S. C. (2016), "Combined Influence of Geometric Imperfections and Misalignment of Journal on the Performance of Four Pocket Hybrid Journal Bearing," *Tribology International*, **97**, pp. 59–70.
18. Tanner, R. I. (1964), "Short Bearing Solution for Pressure Distribution in a non-Newtonian Lubricant," *Journal of Applied Mechanics*, **31**(2), pp. 350–351.
19. Wada, S. and Hayashi, H. (1971), "Hydrodynamic Lubrication of Journal Bearings by Pseudoplastic Lubricants: Part-1, Theoretical Studies," *Bulletin of Japan Society of Mechanical Engineers*, **14**(69), pp. 268–278.
20. Wada, S. and Hayashi, H. (1971), "Hydrodynamic Lubrication of Journal Bearings by Pseudoplastic Lubricants: Part-2, Experimental Studies," *Bulletin of Japan Society of Mechanical Engineers*, **14** (69), pp. 279–286.
21. Tayal, S. P., Sinhasan, R. and Singh, D. V. (1981), "Analysis of Hydrodynamic Journal Bearings with non-Newtonian Power Law Lubricants by the Finite Element Method," *Wear*, **71**, pp. 15–27.
22. Tayal, S. P., Sinhasan, R. and Singh, D. V. (1982), "Finite Element Analysis of Elliptical Bearings Lubricated by a non-Newtonian Fluid," *Wear*, **80**, pp. 71–81.
23. Sinhasan, R. and Goyal, K. C. (1995), "Transient Response of a Two-Lobe Journal Bearing Lubricated with non-Newtonian Lubricant," *Wear*, **28** (4), pp. 233–239.
24. Hayashi, H., Wada, S. and Nakarai, N. (1977), "Hydrodynamic Lubrication of Journal Bearings by Pseudoplastic Lubricants Effects of Composite Flow Characteristics," *Bulletin of JSME*, **20** (140), pp. 224–231.
25. Safar, Z. S. (1979), "Journal Bearings Operating with non-Newtonian Lubricant Films," *Wear*, **53**, pp. 95–100.
26. Raghunandana, K. and Majumdar, B. C. (1999), "Stability of Journal Bearing Systems using non-Newtonian Lubricants: A Non-Linear Transient Analysis," *Tribology International*, **32**, pp. 179–184.
27. Jang, J. Y. and Chang, C. C. (1987), "Adiabatic Solutions for a Misaligned Journal Bearing with non-Newtonian Lubricants," *Tribology International*, **20** (5), pp. 267–275.
28. Jang, J. Y. and Chang, C. C. (1988), "Adiabatic Analysis of Finite Width Journal Bearing with non-Newtonian Lubricants," *Wear*, **122**, pp. 63–75.

29. Javorova, J., Mazdrakova, A., Andonov, I. and Radulescu, A. (2016), "Analysis of HD Journal Bearings Considering Elastic Deformation and non-Newtonian Rabinowitsch Fluid Model," *Tribology in Industry*, **38** (2), pp. 186–196.
30. Bhujappa, N. N. and Mareppa, R. (2013), "Non-Newtonian Effects of Rabinowitsch Fluid on the Performance of Inclined Stepped Composite Bearings," *Tribology Online*, **8** (3), pp. 242–249.
31. Chetti, B. (2018), "Combined Effects of Turbulence and Elastic Deformation on the Performance of a Journal Bearing Lubricated with a Couple Stress Fluid," *Proc IMechE Part J: Journal of Engineering Tribology*, **232** (12), pp. 1597–1603.
32. Hayashi, H. (1991), "Recent Studies on Fluid Film Lubrication with non-Newtonian Lubricants," *JSME International Journal, Series III*, **34** (1), pp. 1–11.
33. Sheeja, D. and Prabhu, B. S. (1992), "Thermal and Non-Newtonian Effects on the Steady State and Dynamic Characteristics of Hydrodynamic Journal Bearings - Theory and Experiments, *Tribology Transactions*, **35** (3), pp. 441–446.
34. Das, S. and Guha, S. K. (2019), "Numerical Analysis of Steady-State Performance of Misaligned Journal Bearings with Turbulent Effect," *Journal of the Brazilian Society of Mechanical Sciences and Engineering*, **41**(81), pp. 1–10.
35. Derdouri, A. and Carreau, P. J. (1989), "Non-Newtonian and Thermal Effects in Journal Bearings", *Tribology Transactions*, **32**(2), pp. 161–169.
36. Hinze, J. O. (1959), "*Turbulence: An Introduction to Its Mechanism and Theory*," McGraw-Hill, New York.
37. Huebner, K. H. (1975), "*The Finite Element Method for Engineers*," John Wiley & Sons, Inc., New York.

# 6 Bio-Functionalized Porous TI

*A. I. Costa*
Universidade do Minho
University of Porto

*J. Géringer*
Université de Lyon

*F. Toptan*
Universidade do Minho
UNESP

**CONTENTS**



## 6.1 INTRODUCTION

A remarkable increase in human life expectancy led to an unprecedented shift in the current patterns of disease, showing an increase in degenerative diseases, which causes degeneration of the bones. More than 50% of chronic diseases in patients more than 65 years old in Europe are related to bones and joints pathology. This brings a growing requirement for implants not only due to the people's longer life but also due to the fact that more people at a younger age are injured from physical activity or trauma [1–4].

In most of the cases, the evolution of degenerative diseases brings prostheses as the sole solution with the demand for artificial bone implants increasing all over the world. Unfortunately, the available implants that are currently used have limited durability, most of the time, due to degradation in service. The products resulting from the degradation processes are bio-reactive species that, by interacting with the human tissues, provoke an adverse reaction that may lead to implant rejection and

DOI: 10.1201/9781003096443-6

failure and consequently, a revision surgery. Nowadays, more hip replacements and dental implants are required for younger or more active patients, mainly due to traumas and diseases, making it essential and mandatory to improve their long-term longevity [2,5–9].

Each year, more than 2.2 million patients undergo a hip or a knee replacement in OECD countries. Since the beginning of the 21st century, hip replacement rates have tripled. Currently, European countries like Germany, Austria, Switzerland, Finland, Luxembourg, and Belgium present the highest rates for hip and knee replacement while Mexico, Portugal, Israel, Ireland, and Korea present the lowest rates. Differences in the population structure may explain part of this variation across countries [1].

Health care quality is achieved by measuring what is important for people. However, just a few health systems in the world are asking and registering patients' outcomes after their surgeries. In particular, after a hip replacement, according to OECD indicators, an individual's quality of life, regarding mobility, self-care, activity, pain, and depression, is improved around 20% [1].

The major clinical problems, illustrated in Figure 6.1, leading to the implant failures are:

- Infections that are caused by infectious agents, mainly bacteria; it is challenging to treat orthopedic implant infections due to a resistance of conventional therapy and this may lead to revision surgeries and in most severe cases, it can result in amputation and mortality. Infections are associated with patient health factors and it was found that aspects like chronic diseases such as rheumatoid arthritis and diabetes, age, and high body mass index are among the main reasons that increased their risks. Infections may also occur when the implant material provides surfaces with vulnerability for attachment and proliferation of bacteria, and consequently to biofilm formation, i.e., adherent bacteria that produce a protective and polymeric extracellular substance become more difficult to be treated. A revision surgery increases the risk of infections [10–13].
- Aseptic loosening of the implant initiated by Young's modulus mismatch existent between the bone (10–30 GPa) and the implant (for example, 110 GPa for Ti) leading to a stress-shielding effect and consequently to bone resorption. In this way, aseptic loosening reported the mechanical failure between the implant and the bone and appears primarily as a consequence of focal periprosthetic inflammatory bone loss occurring at the interface of the implant. At present, aseptic loosening is the cause of more than 50% of the revision surgeries in hip arthroplasty. Specifically, early aseptic loosening may be related to the characteristics of the material and implant design that may affect osseointegration [7,14–18].
- Bioinertness of the implant surface resulting in a poor bond between the surrounding bone tissue and the implant that causes fibrous encapsulation, a layer of connective tissue formed as part of the body's response to a foreign material [7,11,14,19].

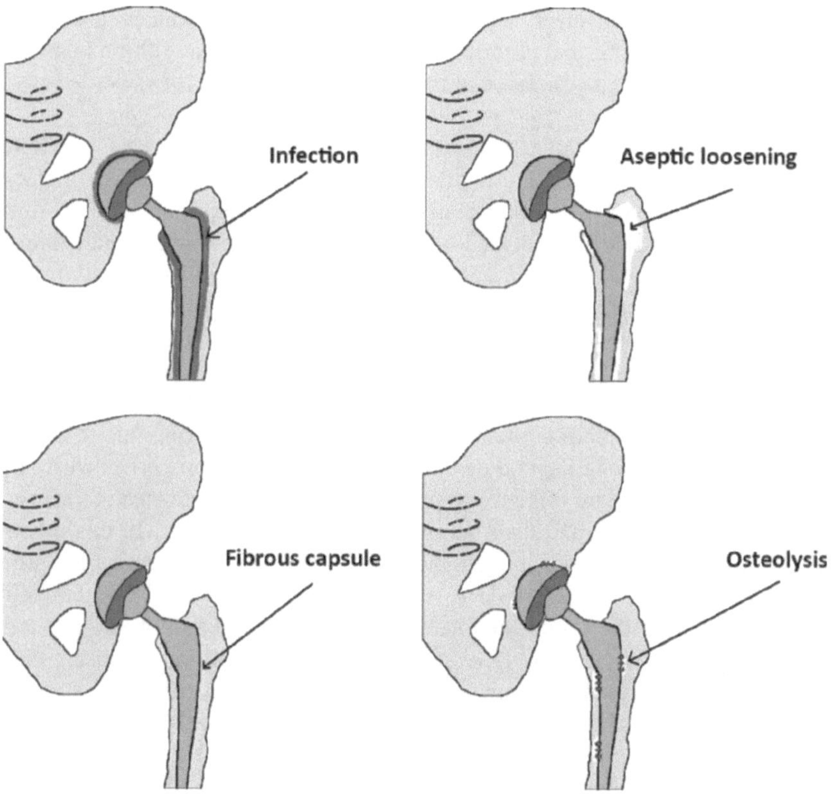


**FIGURE 6.1** Schematic drawing of the major clinical problems leading to hip implant failures.

- Osteolysis due to an inflammatory reaction caused by wear debris released for the surrounding tissue of the implant due to the bio-tribocorrosion process that the material suffers, caused by the synergism of the physiologic aggressive environment together with the tribological wear as a degradation process [2,7,14,20,21].

These problems will prevent the bone to heal and significantly reduce the quality of life of the patients. In this way, new investigations are demanded to improve clinical orthopedic results [22]. Some of the solutions investigated are:

- By controlling the implant macro-porosity, it is possible to adjust the stiffness of the implant to match the one of the bone and consequently, reduce or eliminate the stress-shielding effect caused by the stiffness mismatch between the prosthesis and bone [23–25];
- Through surface modification, it is possible to improve the bioactivity, corrosion, and tribocorrosion performance of the implant by creating a

micro-porous oxide layer with bioactive elements and this can enhance biological mechanisms and improve the long-term performance of implants by improving the wear resistance of the implant [26–28].

In order to succeed, implants require varied mechanical properties in different regions of their structure. The most common examples are dental, hip, craniofacial, and mandible implants. For each application, the functional requirements and esthetic aspects must be considered. For load-bearing implants, like hip and mandible, the mechanical properties change throughout the implant and the functionality requirements of the implant are heterogeneous through the implant. In the case of the hip, the head part should be dense while the stem can be porous. In the case of the mandible, the dental abutment must be a dense region while the body of the implant may be a porous part. These structures are described as Functionally Graded Materials (FGMs). FGMs are attracting attention in biomedical engineering due to the possibility to have exceptionally good mechanical and biological properties where they are most solicited. A more effective implant requires a gradient structure across it, with tailored size, shape, and distribution of pores in each region. In this way, the part with a high level of porosity can stimulate elevated vascularization and direct the formation of bone while the dense structure can provide mechanical stability. Somehow, it mimics the porous architecture of the bone because it presents a differentiated and graded structure [7,19,25,29,30]. Thus, the following aspects should be considered for designing implant surfaces: mimicking the natural bone structures at macro-, micro-, and nano-scales with an adequate chemical composition, wettability, and rapid osseointegration ability. Osseointegration can be defined as the capability to produce bone from the osteoblasts that adhere to the substrates and rapid osseointegration is a critical factor for the success rate of orthopedic implants. Osseointegration is a crucial stage for the success of implants and it has been shown that the biological fixation is closely related to the surface characteristics of the implant material. In this way, in order to fulfill such characteristics to achieve better outcomes, different materials together or not with coatings and surface modifications have been suggested to enhance the biomechanical behavior in order to increase the performance of the implant–bone interface [22,31–34].

A biomaterial can be defined as a material designed to take a form that can direct, through interactions with living systems, the course of any therapeutic or diagnostic procedure [35]. Its production for implantology requires a multifaceted team composed of physicians, scientists, and engineers who get involved in the creation of a new medical device in order to increase the quality of life of the patients. The development of new materials is constantly aiming for the optimum balance between the biological, mechanical, and triboelectrochemical performance as well as the surface properties since they can influence biological and mechanical compatibility. So, the desire is to develop biomaterials with mechanical properties that match the bone, without compromising its biocompatibility and bioactivity [14,25,36]. Biocompatibility can be defined as the ability of a material to perform with an appropriate host response in a specific application and it is controlled by the initial and successive reactions between the material and host body, where the most important are related to adsorption of molecules and proteins, cell and bacterial adhesion, activation of macrophage, formation of tissues and inflammation processes [37]. In this way, biocompatibility should be considered a characteristic of a biomaterial–host interface and not just from the biomaterial itself [38].

Several orthopedic clinic applications, for example, knee, hip and shoulder joint implants, bone plates, and screws, among others, have been used in different biomaterials with the aim to improve performance for such applications. Currently, almost 80% of implants used are made of metallic biomaterials and the most metals used are Ti and its alloys, stainless steels, and cobalt (Co)–chromium (Cr)–molybdenum (Mo) alloys. Among these, Ti exhibits the highest biocompatibility, corrosion resistance, and specific strength (ratio of tensile strength to density). Furthermore, metallic biomaterials, such as magnesium (Mg) alloys, iron (Fe), tantalum (Ta), and niobium (Nb), are also used in the field but with a much smaller presence [39].

One of the most abundant metals on Earth's crust and lithosphere is Ti, a transition metal within group IV of Mendeleev's periodic table. Elements from this group present a partially filled d subshell and they are very reactive in the presence of oxygen and that is why an oxide (passive) film is immediately formed on their surfaces. Ti presents the highest strength-to-weight ratio between the metals group. Ti is abundant and widely distributed in natural mineral deposits (ilmenite and rutile) making it more reachable than rare elements [22,32,40,41]. Ti is not usually used in the automotive industry due to the high cost of extraction and production. Therefore, Ti finds applications mainly in sectors with demands, such as the aerospace industry or biomedical devices, where the final high cost is not the principal concern [42]. More than 1000 tons of Ti are implanted into patients every year. While Ti implants may have achieved high success rates, two major problems are still experienced: the lack of bone tissue integration and implant-centered infection. Ti is susceptible to crack propagation, which, together with the other clinical problems associated with the implant failures, can limit the lifetime of Ti implants to only 10–15 years [7–9,15,25,40,43,44]. Besides this, implants made of metallic materials, including Ti, are much stiffer when compared to bone, and this mismatch can cause stress-shielding effects under load-bearing leading to bone resorption and consequently, an eventual failure of implants. Therefore, metallic biomaterials with lower Young's modulus that can match the one from the bone are required in order to minimize the stress-shielding effects and to guarantee an optimal load transfer [17]. Among these, Ti and its alloys have been used extensively and there are studies on developing low Young's modulus implants, but they still could not reach the modulus of the bone [45–47]. Although new β-type alloys are recently emerging for biomedical applications with promising properties mainly in terms of their mechanical behavior, there is still a lack of exhaustedly completed studies, especially under in vivo evaluation.

In this way, another route of investigation that had progressively shifted the focus from an interface dense metallic bone–implant design is porous metallic structures that may present a Young's modulus that fits the one found in the bone. Besides this, a porous structure can promote a mechanical interlocking and more area to contact with cells, leading to a bone ingrowth inside of the pores, as shown in Figure 6.2 [48–53]. Porous Ti maintains the already known properties of a material with the possibility of osseointegration inside of the pores. This porosity gives the possibility

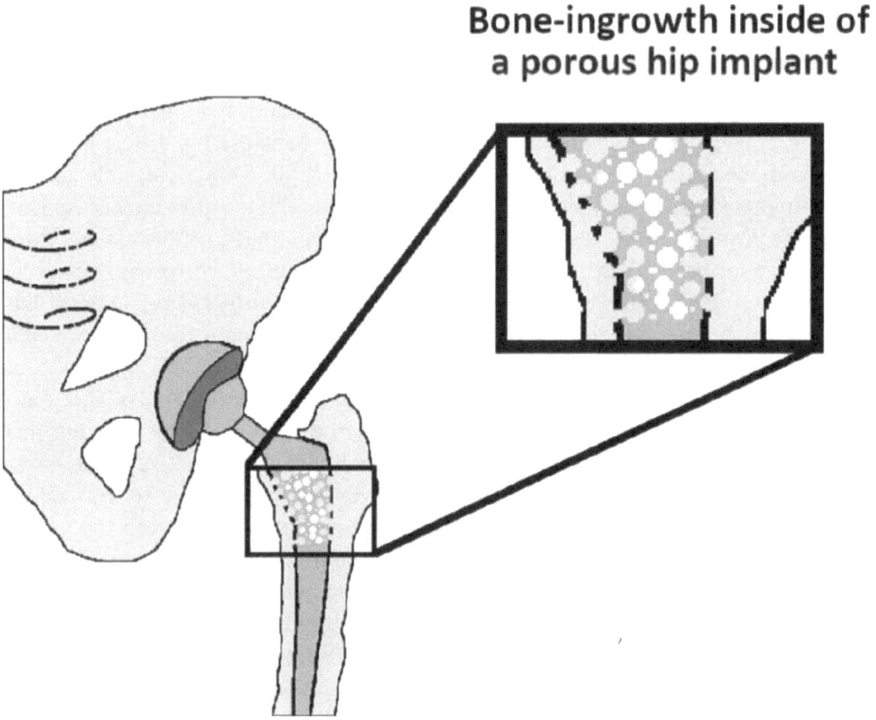


FIGURE 6.2 Schematic drawing of bone ingrowth inside of a porous hip implant.

of development of materials in accordance with the demands of specific situations encountered in clinical practice. While factors controlling their biological and mechanical response have recently been a subject of research [48–52], the interplay between mechanical and triboelectrochemical properties, bone-ingrowth requirements, and manufacturing constraints is still under study [26,54,55]. The processing of an implant with hierarchical porosity with macro- and micro-properties will determine its performance since they will control the physicochemical characteristics and topographical parameters that are the main communication with cells [56].

## 6.2 POROUS TITANIUM

With porosity, it is possible to obtain maintenance of the mechanical strength of the implant and at the same time provide the adequate pore size for bone ingrowth. Porosity is the best way to achieve the required weight and density and it determines the final mechanical properties of the implants. Porosity may be defined by the percentage of void space in a solid and is a morphological property completely independent of the material. Pores provide the cells with the necessary space to migrate, to allow vascularization and proliferation, and eventually to differentiate into the needed cell. Depending on the percentage of porosity and pore geometry, the surface area available to cell adhesion will be different and that will influence the vascularization process. In order to ensure ingrowth of bone inside of the pores, the optimum porosity may vary from about 30% to 60% [22,29,30,40,50,57].

A porous structure induces less time for the osseointegration process because it favors the contribution of adequate nutrients for the osteoblasts. Open pores in the porous structures are preferred for the increased ingrowth of tissue inside of the porous structure and cells on the surface of the material [36]. A complex structure is required with a wide range of pore sizes. Macro-pores of 50–300 μm are required for the fixation of the implants due to the interlocking between the tissue that grows inside of the pores [57,58]. However, an optimum range for the pore size required for implant fixation and osseointegration remains undetermined. From a biological point of view, some authors [30] prefer to consider a larger range and stated that an open porous structure with pore sizes from 50 to 800 μm should be the aim in order to favor bone ingrowth. On the other hand, other authors [40,59] follow a smaller range and consider that pore sizes between 100 and 500 μm are the most suitable for bone ingrowth and humoral transmission, whereas the minimum pore size for acceptable bone ingrowth may be in the range of 50–150 μm [57]. Baril et al. [60] used micro-computed tomography in order to quantify bone ingrowth inside of porous Ti. With this technique, the authors visualized and were able to quantify the bone amount in different implant regions and planes and showed that bone ingrowth near cortical bone was around 28.9%, significantly greater than that in cancellous that was found to be about 14.5%. The authors also presented mineralized tissue inside of the porous structures and stated that a pore size larger than 28.5 μm did not affect bone formation, suggesting 28.5 μm as the minimum pore size for efficient cell migration.

From a biological point of view, after the implantation, the response can be of two types. The worst type is related to the formation of a fibrous soft tissue capsule around the implant that does not provide biomechanical fixation and consequently leads to the clinical failure of the implant. On the other hand, a better response involves the direct bone–implant contact, without having a connective tissue layer. The response is largely influenced by the surface chemical composition, hydrophilicity, and roughness since such parameters play an important role in implant–tissue interaction and osseointegration. Both bone anchoring and biomechanical stability are improved with implants with rough and porous surfaces [43,61,62]. Porosity, roughness, and wettability are considered three major factors that contribute to the success of a cell-contacting biomaterial since the contacted area between the cells and the material is enlarged and in this way, cell growth is promoted [63]. Although it was accepted that a rough surface is considered beneficial for cell adhesion and proliferation, the interplay between roughness and mechanical properties needs to be taken into account, especially for porous structures. Wang et al. [64] showed that the best etching treatment in porous Ti for cell adhesion and proliferation presented a negative effect on the mechanical properties of such structures.

The production of porous Ti must provide high manufacturing rates and competitive prices so it is very important that porous structures present economically viable fabrication methods. Porous Ti implants will demand technologies that can concede economic mass production for standard implants, while preserving flexible manufacturing options for costume-made implants and this market will sharply rise [25].

There are several ways to produce porous Ti structures. Powder metallurgy (PM) [26,49,55,65,66], additive manufacturing (AM) [29,67–70], freeze-casting [71,72], spark plasma sintering [73,74], and metal injection molding [75,76] are some of the available techniques to produce porous structures. The high melting point of Ti (about 1670°C) combined with its high reactivity makes liquid-based techniques not viable. Therefore, the production of porous Ti through powder-based processing routes is considered an economic and flexible way to achieve the desired designed structures. PM presents several advantages such as high material yield, reduction of machining steps, and the possibility of designing alloys with tailored compositions. In addition, some of the above-mentioned methods provide limited porosity. PM can be combined with the space holder technique and in this way, it is possible to obtain uniformity, adjustable porosity level, controlled pore shape, and uniform pore size distribution with a not-so-expensive technology. Sintering temperature, porosity, and pore size will condition yield strength and Young's modulus [24,49,55,59,65]. Several space holders can be used like urea or carbamide ($CH_4N_2O$) [24,26], sodium chloride (NaCl) [49], ammonium hydrogen carbonate $(NH_4)HCO_3$ [66], acrowax [66], sugar pellets ($C_{12}H_{22}O_{11}$) [77], tapioca [78], and rice husk [79].

AM offers freedom to design and a wide choice of materials and even multiple materials. Recently, AM techniques showed flexibility in constructing micro- and macro-scale structures based on computer models and analyses. AM can also be used for near-net-shape fabrication of complex-shaped parts with tailored properties [29,67–70]. Nowadays, AM shows promising techniques to produce porous Ti structures, between them, the most common ones are electron beam melting (EBM) [29,69,70] and selective laser melting (SLM) [80–82]. EBM is a rapid manufacturing process in which an electron beam energy is applied on overlaying materials layer by layer based on CAD data. During the process, the electron beam passes over each successive layer of powder, which is gravity-fed from powder cassettes and raked into successive layers. The SLM technique produces by melting selected areas of powder layers using a computer-controlled laser beam. Both techniques provide complex porous structures with freeform of shapes and porous structures with controlled pore size fraction. Usually, the properties achieved are comparable to the design parameters but the predictability at higher porosities seems to be better [29,69,70,80–82].

Impurities can seriously affect the mechanical properties of porous Ti, where the amounts of oxygen and nitrogen play a crucial role. Increasing oxygen content develops an increase of the yield strength, hardness, and fatigue resistance at a given stress level, whereas on the other hand, leads to a decrease of the ductility and impact resistance by restricting twinning and prismatic slip. When heated to a high temperature, Ti derives its strength from solution hardening because of oxygen and nitrogen and the presence of these interstitial elements leads to the brittleness of the material [40,83,84].

The oxygen content has an important impact on the compressive properties of dense and porous Ti. The yield strength and the ductility are significantly affected when the oxygen content increases from 0.24 to 0.51 wt%. After the sintering stage in a vacuum, the surface of Ti is repassivated when exposing to air and this is an additional oxygen pick-up. This effect is minimal for dense components but may be very relevant for porous components. This oxygen film contributes to the total oxygen content but its influence on the final mechanical properties is usually negligible. However, this oxide film provides improved corrosion resistance and biocompatibility to Ti [83–85].

The use of particles in most of the processes for the production of porous Ti provides a large surface area that is naturally covered with a thin oxide film that contributes to an increase in the oxygen content and consequently can affect the properties of the final materials. The large surface area of the powder provides also a large portion of the material to react with additives (i.e., binders and space holders) or with the atmosphere during processing. In addition, the high-temperature processing of the powder leads to a dissolution of the oxide and/or nitride layers on the surface of the powder and, therefore, provides a fresh Ti surface to react with the atmospheric constituents. Then, a natural oxide layer reforms on the surface of the porous structure after exposure to air at low temperatures. While the impact of interstitials on the properties of dense Ti-based materials has been widely studied and reported in the literature, their effect on the properties of porous Ti structures has been rarely reported [85,86].

Implants in the human body are usually under cyclic loading conditions during walking and running. The cortical bone suffers combined loading stress states that can involve tension, compression, shear, and fatigue [87]. There have been several studies evaluating the compression behavior [50,59,87–89], bending behavior [90,91], and fatigue behavior [84,87,92,93] of porous Ti. However, final results may present variations depending on the details of processing techniques, interstitial elements content, and heat treatments, among other details.

The mechanical properties, especially Young's modulus, of porous Ti are mainly governed by the porosity. Xiang et al. [59] showed that the compressive strength and Young's modulus decreased as the porosity level increases. When the porosity was 70%, Young's modulus was 1.2 GPa, which is close to the one found in the human bone. Imwinkelried [90] stated that the typical yield strength of the porous Ti with 62.5% porosity is above 60 MPa in compression, bending, and tension. In conclusion, Young's modulus, the yield strength, and the ultimate compressive strength decrease with an increase in porosity [40]. Nevertheless, the choice of the testing method for the evaluation of Young's modulus of porous materials is critical. The most-reported technique is the compression test; however, dynamic methods like the ultrasound technique have superior precision and repeatability. Hence, the literature shows that Young's modulus measurements from uniaxial compression tests are significantly lower than those obtained by dynamic measurements [94,95].

Bending combines tension and compression in a more complex stress distribution so that is why the values obtained by the bending test differ from the ones obtained in pure tension or pure compression [96]. Amigó et al. [91] tested the bending performance of porous Ti produced by PM with the space holder technique and stated that the size of the space holder particles influences flexure strength. More specifically, a higher bending strength was obtained on the porous samples that had smaller particle sizes of the space holder. The results also showed that bending strength decreased as porosity increased. The same conclusion was found by Kashef et al. in their study [97], where lower porosity provided a higher bending yield stress.

Fatigue may be defined by a progressive localized permanent structural change occurring in a material subjected to cyclic stresses that are often the principal cause of the premature mechanical failure of the metallic biomaterials. Under fatigue conditions, in the human body, fluids can accelerate the initiation of a surface flaw and can lead to propagation to its critical size that can start the fracture. The simultaneous action of fatigue stress and electrochemical dissolution is defined as corrosion fatigue. Fatigue strength presents remarkable changes when studied under corrosive action of simulated body fluids [87,92,98–100]. The reduction of fatigue life of Ti and its alloys under corrosion fatigue has been scarcely documented [101–105]. However, especially for porous structures, where it is possible to have interconnected porosity allowing fluids to penetrate through the structure, corrosion fatigue must be explored. Figure 6.3 presents surface and cross-section tomographic slice images

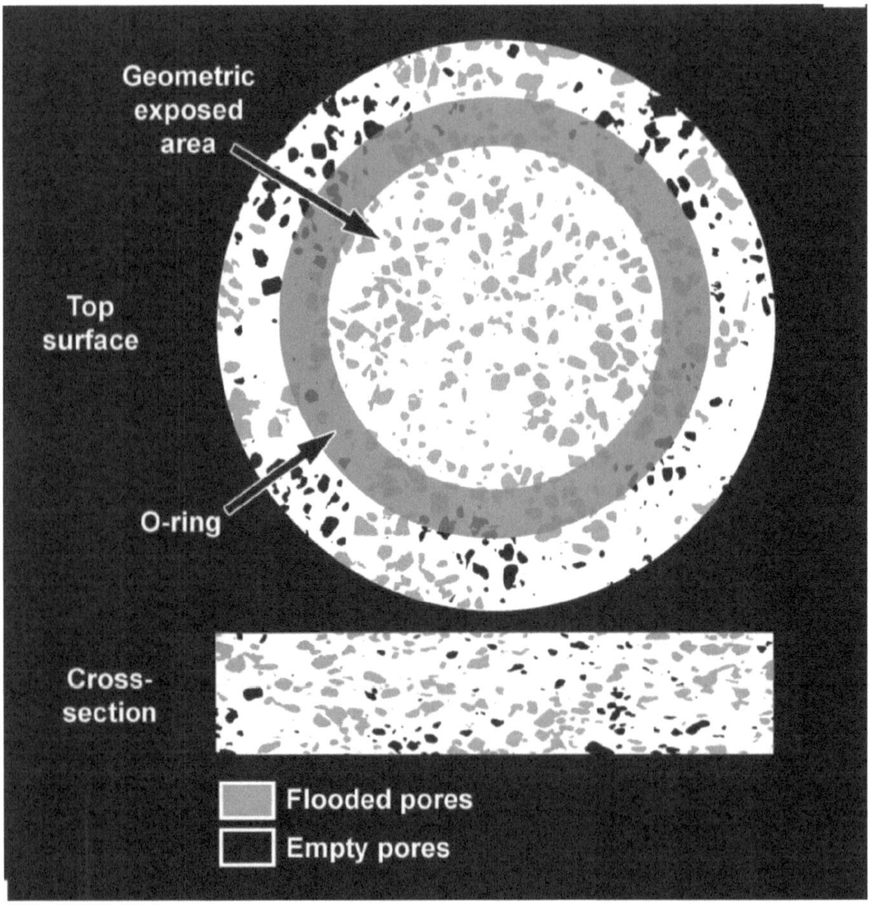


**FIGURE 6.3** Top surface and cross-section tomographic slice images presenting a simulation of electrolyte penetrating the pores. (Reproduced with permission from Ref. [26], Copyright (2020).)

showing a simulation of flooded pores by electrolyte, where porous Ti with a mixture of open and closed pores was found to be crossed by the electrolyte, meaning that the top and bottom surfaces were inked by a void structure [26].

Porosity may be an efficient way to achieve the mechanical and biological characteristics that are needed; however, it may present a detrimental effect on the corrosion and tribocorrosion properties. The lifetime of an implant is crucial and it is conditioned by its degradation and this process is responsible for decreasing the structural integrity of the implant and the release of wear and corrosion products that may initiate some adverse biological reaction in the body and even lead to mechanical failure of the implant. Body fluids present a highly aggressive chemical environment for the metals due to their high concentration of chloride ions that have the ability to induce localized corrosion [106–109]. The study of the electrochemical behavior of Ti is well extended, whereas the same study on porous Ti is scarce. In general, studies indicate that pores may have strong effects on corrosion characteristics [110–112]. Xu et al. [110] fabricated porous NiTi and presented a decrease of the corrosion resistance on the porous samples that was attributed to a larger real contact surface area of the porous structures, as well as, the presence of crevice corrosion. Alves et al. [113] studied the corrosion behavior of porous Ti having 30 and 50 vol.% of nominal porosity. Electrochemical studies revealed that samples presented a less stable oxide film at the inner pores at increased porosity. The lack of a well-defined passivation plateau in the potentiodynamic polarization curves was explained by the heterogeneities on the oxide film formed, especially on the most inner pores. The authors stated that the difference in the thickness and the nature of the oxide film between the porous structures was due to the difficulty in the electrolyte penetration through the pores as well as the difficulty in the oxygen diffusion. More recently, Alves et al. [26] developed corrosion studies on macro-porous Ti and stated that the corrosion rate of macro-porous structures was influenced by the surface area.

Tribocorrosion is the phenomena caused by both the electrochemical and mechanical interactions on the materials, i.e., synergistic combination of corrosion processes and wear is one of the most important aspects in biomedical industries. It appears as a threat to the quality of implant systems during and after the implantation surgery. There has been significant progress in this research area; however, studies are significantly less when compared to the ones related to tribology and corrosion areas, separately.

Hip implants as well as dental implants suffer different kinds of tribocorrosion phenomena. There are different interfaces in the hip implants: bone-cup (fretting corrosion), head-cup (sliding tribocorrosion), head-neck (modular junction: fretting corrosion), and stem-bone (modular junction: fretting-corrosion), as presented in Figure 6.4. Dental implants are exposed to cyclic micro-movements at the implant–abutment interface causing relative motion between contacting surfaces and leading to wear [114–116]. The wear behavior of the implant depends on several properties, among them, with special emphasis on hardness, roughness, fracture toughness, and Young's modulus of the interacting materials. In this way, implants must have a good performance regarding degradation that will not interfere in the bone–implant interface [117].

Porous structures may present different wear behavior compared with dense materials. Studies [55,118–120] showed that porosity can play either a favorable or

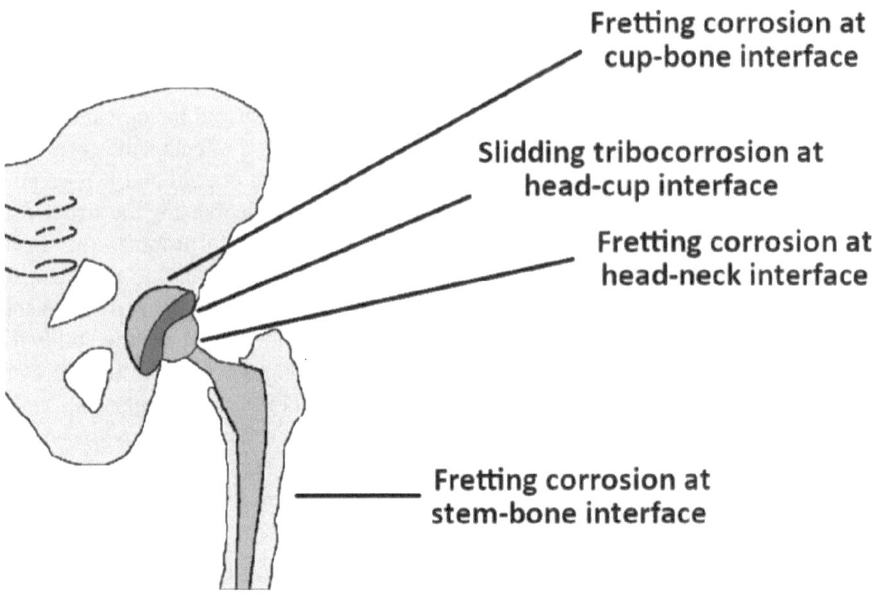


**FIGURE 6.4** Schematic drawing showing the fretting corrosion and sliding tribocorrosion interfaces in the hip implant.

an unfavorable role. Pores at the surface reduce the real nominal contact area and may trap the wear debris where they can be partly compacted in the pores. The presence or absence of debris can influence wear mechanisms and the coefficient of friction. Porosity can also contribute to the initiation of microstructural defects, such as cracks that may produce higher wear rates and decrease the transition load. Toptan et al. [55] studied the tribocorrosion behavior of macro-porous Ti and stated that the ejection of the wear debris into the macro-pores can decrease the third-body abrasion effect and may contribute to an improved tribocorrosion performance by macro-pore structures. A schematic diagram of the suggested tribocorrosion mechanisms for porous Ti is presented in Figure 6.5.

In contrast to what happens in static corrosion resistance, where the performance depends on a stable spontaneous passive film that was formed, under tribocorrosion evaluation, the behavior relies on depassivation of the passive film and the formation of a tribofilm (including adsorbed proteins) [45,121,122]. It has been reported [5,123] that proteins affected the formation and the growth of the passive film and changed the variation in corrosion kinetics. Biological products (e.g., cells, proteins) can be a part of an extra layer that protects the implant from wear and, therefore, may decrease the damage caused by tribocorrosion.

Thus, the use of porous Ti instead of a fully dense material may allow primary stability for the implant and a possibility for the living bone to grow inside of the structure. Although having these advantages, Ti is presented as a bioinert material, and after the implantation, Ti does not bond to the bone directly. Therefore, it is desirable to perform changes in the surfaces in order to add some bio-function without

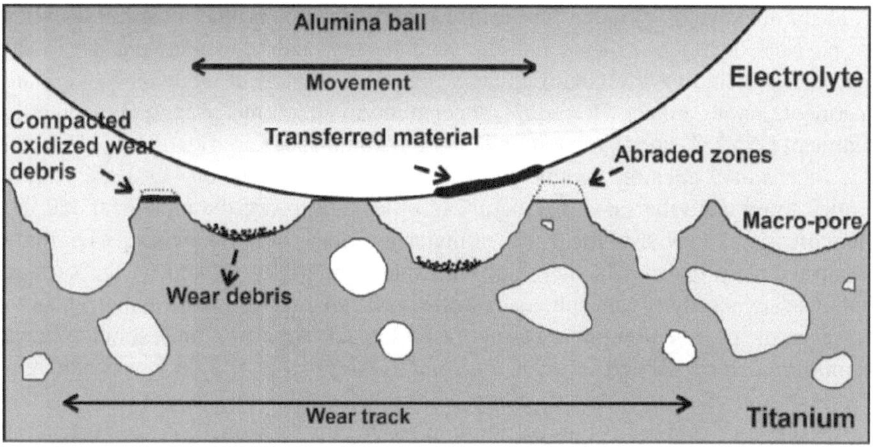


**FIGURE 6.5** Schematic drawing for a possible wear scenario for porous Ti.

compromising the stable fixation. It is also pertinent to do improvement against long-term concerns such as the generation of metal debris from the implant. With surface modification, it is possible to achieve bioactivity, enhanced by the incorporation of bioactive elements, and to achieve protection from both wear and corrosion [43,117,124].

## 6.3 BIO-FUNCTIONALIZATION BY MICRO-ARC OXIDATION

Ti presents low resistance to wear and a high coefficient of friction, releasing metal ions under tribocorrosion solicitations [27,123,125,126]. However, there is still no consensus on how much amount of metal ions can be tolerable in the human body [127]. Recently, Costa et al. [128] showed that particles from tribological tests performed on Ti alloys increase the production of anti-inflammatory cytokines. Wear particles from Ti-based materials also exhibited the capability to penetrate osteoblast cell membranes and to become internalized [128]. In this way, the use of an appropriate surface modification is required to improve the surface properties of the implant [14,20,21,27,125].

Ti implants are expected to detain high bone–implant contact. Briefly, there are four factors that can be considered to be relevant in the contact between the bone and the structure: materials and surface treatments, topography, surface energy, and bio-functionalization [32,129].

The main disadvantage of using metals, like Ti, as biomaterials is the fact that they have no bio-function. In order to solve this issue, surface modification is an effective way to bio-functionalize, altering the surface morphology and chemical composition of metals to be used as biomaterials where the tissue compatibility of the surface can be improved. Surface modification with nano- and micro-porous structures showed promising results at the implant–bone interface, with improved cell response, improved corrosion and wear resistance, and, consequently, an increased implant lifetime [22,130–133].

There are several methods for surface modifications, and the most known ones are physical techniques like polishing, sandblasting, and heat treatment, chemical approaches including acid/alkali etching, and electrochemical methods like anodic treatment, among others. Nowadays, surface modification methods, such as anodic treatment [125,134], ion implantation [135], nitriding [135], laser-cladding [135], surface mechanical attrition treatment [136], and thermal oxidation [136] are widely applied to regulate the surface structure in order to improve the biological and the tribocorrosion behavior of the Ti-based materials. Some of those techniques enforce the protective properties of the oxide film that is naturally formed on the surface that consists mainly of amorphous or low-crystalline and non-stoichiometric $TiO_2$. However, the passive film possesses poor tribological properties; thus, some of those techniques can be applied in order to promote the growth of a thick and adherent oxide layer on the surface and to enhance its tribocorrosion resistance [137].

At least four different types of titanium oxide can be formed on the surface of Ti. Under atmospheric pressure, the thermodynamically stable oxide is $TiO_2$ and it can exist in an amorphous natural state or in one of the three crystalline polymorph phases: anatase (tetragonal), rutile (tetragonal), and brookite (orthorhombic). Rutile is the most stable crystallographic phase of $TiO_2$, but lower temperatures favor the formation of metastable anatase over the rutile [137].

Among the various available anodic treatments, micro-arc oxidation (MAO), which can also be called plasma electrolytic oxidation (PEO), generates a porous uniform hard $TiO_2$ layer that it is firmly adherent to the bulk material (Ti). This $TiO_2$ layer already proved to be a promising approach since it is capable of improving the surface aspects of materials intended for biomedical applications by presenting better roughness and wettability and promoting bioactivity. Consequently, these modifications bring better results on cells adhesion, proliferation and differentiation, blood compatibility, and reduction of the hemolysis rate, among other properties [28,48,138,139]. In addition to improved bioactivity, many other studies refer to MAO as a suitable and economical method for improving the corrosion [140–145] and tribocorrosion [55,125,126,146,147] behavior because the hard $TiO_2$ layer acts as a physical barrier against corrosion and wear.

MAO provides the possibility of combining the chemical and morphological modification of Ti implant surfaces in an effective single-step environmentally friendly and fast process. Various electrolyte solutions are used to anodize Ti and the composition of the new layer is mainly determined by the electrolyte constituents [33,148,149].

There are several biological studies on anodized Ti surfaces [134,150,151]. Chen et al. [150] successfully applied MAO in order to grow $TiO_2$ (anatase- and rutile-rich) layers on β-Ti alloy for improved osseointegration. In vitro osteoblast activity and microscopic observations revealed that osteoblasts grew well and completely fused with the micro-porous $TiO_2$ layer. Oliveira et al. [134] reported that the anodic treatment with CaP incorporation modulated positively the cell viability and osteoblast-related gene expression [152]. Ribeiro et al. [28] studied MAO layers formed on Ti surfaces and showed faster osteoblast adhesion and spreading, when compared with Ti. Felgueiras et al. [151] demonstrated that porous oxide surfaces promoted osteoblast attachment, differentiation, and consequently osseointegration.

These structures allowed a stronger bone anchorage in both animal and human experiments. The authors performed tribocorrosion tests in the presence of MG63 osteoblast-like cells and the results showed a resistant oxide layer and cells' capacity of regeneration after implant degradation [151].

Durdu and Usta [153] studied the tribocorrosion behavior of Ti6Al4V alloy modified by MAO and stated that the wear resistance and tribological properties of the treated samples were better than those of the untreated alloys. Alves et al. [125] stated that the presence of rutile decreased the mechanical damage after sliding so the crystalline structure of the layer formed in the MAO process affected the tribocorrosion behavior. In conclusion, several works that studied the tribocorrosion behavior of the porous oxide layer on Ti [125,151,154] showed that MAO contributes to an improvement of biomedical materials performance in the tribocorrosion field.

In the electrochemical field, studies [140–144] have shown that micro-porous oxide layers produced by MAO demonstrated better corrosion behavior when compared to Ti. Park et al. [140] evaluated the surface characteristics of Ti modified by MAO and reported an increased corrosion potential and decreased corrosion current density. Velten et al. [142] reported that the thickness of the $TiO_2$ layer increased with the increase of the oxidation voltage. Thus, a higher thickness of the oxide porous layer can increase the corrosion performance as well as the surface roughness.

There are only a few reports [155–157] regarding the mechanical performance of MAO-treated Ti and its alloys. Gu et al. [156] evaluated the micro-hardness of treated Ti6Al4V samples by MAO and presented a higher value for the treated samples when compared with the untreated ones. On the other hand, Santos et al. [157] studied the mechanical properties of MAO-treated Ti samples by nanoindentation and showed that the hardness of the oxide layer was reported to be similar to the ones found in the Ti bulk material. In contrast, Young's modulus of the oxide layer was 19% lower than the one for bulk Ti due to the porosity.

Hierarchical multiscale (macro-, micro-, and nano-scale) porosity on a surface can be beneficial for promoting cell adhesion and proliferation and for mechanical interlocking [58]. Such an approach can be easily obtained by changing the variables of two or more steps of anodic treatment. Li et al. [58] created a super hydrophilic surface through a two-step MAO process. The authors stated that the first step only changed the morphology, but the second step changed both the morphology and the composition. The authors presented a triple hierarchical structure on the Ti surface, in the macro- (macro-pores with 100–300 μm), micro- (micro-slots with 3–10 μm), and nano- (submicron/nano-pores with 80–200 nm) scale. It was also stated that the ability to induce hydroxyapatite increased due to the presence of the micro-slots and the nano-pores [58]. Zhou et al. [158] performed a three-step MAO on Ti with the incorporation of bioactive elements and biological results presented that good induction of apatite was closely related to the gouges, the structure that was formed in the second step of MAO. More recently, a hierarchical bio-functionalized porous surface was obtained on Ti by a two-step anodic treatment [27] where the authors showed that with the first step, it was possible to obtain macro-porosity while with the second step, a bio-functionalization process by MAO, it was provided an oxide layer with micro-pores and bioactive elements, as presented in Figure 6.6. Corrosion and tribocorrosion behaviors were evaluated and the results showed improved performance

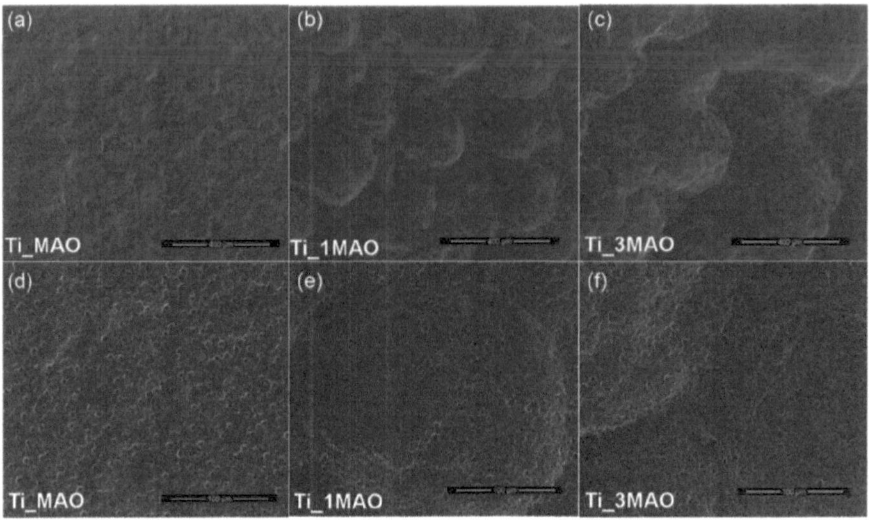


**FIGURE 6.6** Hierarchical multiscale (macro-, micro-, and nano-scale) porosity obtained by the MAO process (a and d) and by a two-step anodic treatment (b–f). (Reproduced with permission from Ref. [27], Copyright (2020).)

due to the protective role given by the hard porous $TiO_2$ layer formed on the second step (MAO).

One of the many advantages of MAO is the possibility of doping bioactive elements to the porous oxide layer with the aim to improve the biological response [28,48, 148, 159–166]. Calcium (Ca) is the most common mineral in the human body, the major structural component of bones and teeth. More than half of the mass of bone mineral is composed of phosphorus (P), which combined with calcium forms hydroxyapatite crystals. Calcium phosphate plays a relevant role to enhance early-stage bone tissue integration [106,167,168]. Ishizawa and Ogino were the forerunners [169] to produce a bioactive oxide layer containing Ca and P, as bioactive elements.

Strontium (Sr) as a bioactive element presents a dual action since it showed beneficial effects on bone growth and reduction of bone resorption. In this way, Sr is reported to increase osteoblast differentiation and bone matrix mineralization [32,148,159–164]. Mg ions naturally occur within the human body and play a critical role in many essential cellular functions, such as activating adenosine triphosphate and synthesizing DNA and RNA. More than 60% of Mg in the human body is found in bone. Mg can be absorbed by the human body over time and this characteristic makes it particularly well-suited for applications that require malleable implant devices [22,159]. Another promising bioactive element is zinc (Zn) that presents a beneficial effect on bone growth and positively affects bone regeneration as well as antibacterial potential [159,170]. Fluor (F) is stored in teeth and bones and increases their structural stability. More recently, F incorporation as bioactive elements showed promising results since it enhanced the antibacterial activity of medical devices and also osteoblastic activity [168]. Silver (Ag) and copper (Cu) exhibit antimicrobial ability; however, their incorporation may lead to cytotoxicity [164,165,170,171]. Recently, it was reported that manganese (Mn) can have a significant role as a co-factor in bone mineralization as well as bone cartilage and collagen formation showing promising results to be used as a bioactive element [36]. Cerium (Ce) is a rare earth element that also has been used in biomaterials to stimulate antibacterial activity [166]. Co has been shown to be favorable for angiogenesis but also exhibits prolonged antibacterial effects [168]. Bioactive elements not only may play a role in biological effects but also can set a performance on the tribocorrosion behavior, playing or not a lubricant effect [172].

## 6.4 BIO-FUNCTIONALIZED MACRO-POROUS TI

The main goal with the bio-functionalization process by surface modification using MAO for porous Ti is to achieve an efficient procedure that provides bioactivity for both the outer surface and inner parts of the surface implant as illustrated in Figure 6.7. Such modification can increase the biological response and tribocorrosion without an alteration on its macro-pore size distribution, amount of porosity, or 3D interconnection. However, the evaluation of bio-functionalized macro-porous Ti is still scarce in the literature [26,48,55,173–175], and most of the studies give a limited evaluation.

Bio-functionalized macro-porous Ti showed apatite-forming abilities with apatite rapidly nucleating and growing on the porous structures [174,175]. Extensive biological studies have been recently reported. Alves et al. [48] studied the biological response in terms of adhesion, spreading, viability, and proliferation of pre-osteoblasts on bio-functionalized highly porous Ti. Representative lower and higher magnification SEM images of osteoblasts attached to bio-functionalized porous Ti are presented in Figure 6.8. The authors stated that bio-functionalization by MAO is an efficient way to incorporate bioactive elements on both outmost surfaces and inner pore surfaces, which may allow the supply of bioactive elements when necessary. Results showed that bio-functionalized porous Ti structures exhibited better cell adhesion and proliferation. Another study [170] with porous Ti implants that were bio-functionalized by MAO showed that bioactive elements such as Ag and Zn were released from the implant surfaces for at least 28 days resulting in antibacterial activity. Besides this,

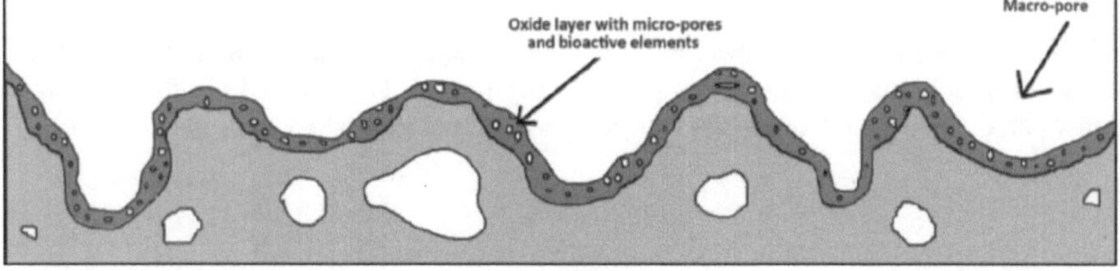


**FIGURE 6.7** Schematic drawing of bio-functionalization of macro-porous Ti, where a micro-porous oxide layer with bioactive elements covers all the surface of the macro-pores.

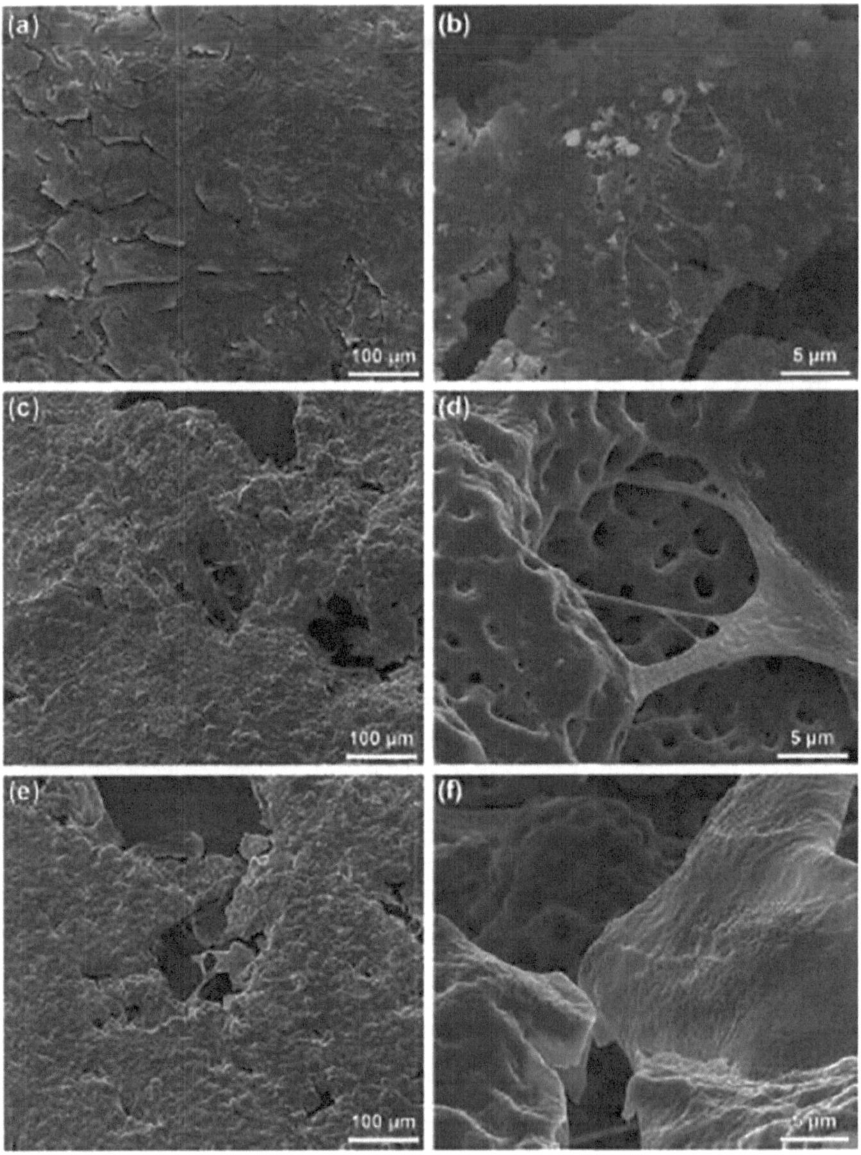


**FIGURE 6.8** Representative lower (a, c, and e) and higher (b, d, and f) magnification SEM images of osteoblasts attached to bio-functionalized dense Ti (a, b) and bio-functionalized highly porous Ti (c–f). (Reproduced with permission from Ref. [48], Copyright (2019).)

bio-functionalized porous Ti generated reactive oxygen species and, by this way, promoted antibacterial contact killing. Such surfaces also showed enhanced osteogenic behavior of pre-osteoblasts after 3, 7, and 11 days [170].

Using a different bio-functionalization process than MAO, macro-porous Ti with $TiO_2$ nanotubes showed potential on the number of planktonic bacteria and hindering the formation of biofilms making them, according to Yavari et al. [165], promising candidates for combating peri-operative implant-associated infections. Bose et al. [167] also studied the effect of $TiO_2$ nanotubes on porous Ti and stated that such an approach induced early-stage osteogenesis with enhanced defect healing and mechanical interlocking at the bone–implant interface.

On the other hand, surface modification methods should not deteriorate the final properties of macro-porous structures. Previously, surface techniques to improve the biological response of porous Ti showed promising results for cell response but the mechanical response was seriously affected [176,177]. Korkmaz et al. [155] applied MAO to open-cell porous Ti6A14V alloy and studied the mechanical properties by compression tests and stated that results of the MAO-treated porous samples were significantly improved when compared to the untreated porous materials, showing higher peak stress. Karaji et al. [178] studied the effect of the MAO process on the mechanical properties of porous Ti processed by AM. The authors showed that when MAO was performed for 2 minutes, it did not influence most of the mechanical properties under studies like maximum stress, yield stress, plateau stress, energy absorption, and elastic modulus. But when the MAO treatment was applied for 5 minutes, a decrease in elastic modulus, yield stress, and plateau stress was shown although no changes were found on the maximum stress and energy absorption.

The triboelectrochemical response is also an important aspect that will condition the overall response of the material that will be implanted. The literature is still very scarce on the evaluation of corrosion and tribocorrosion of bio-functionalized porous Ti. Recently, Alves et al. [26] demonstrated that the bio-functionalization process by MAO improved the corrosion behavior of macro-porous Ti, in terms of thermodynamics and kinetics. The authors reported that an increase in the macro-porosity led to an increase in the corrosion rate, mostly due to the growth of the barrier film formed during MAO. Toptan et al. [55] evaluated the tribocorrosion behavior of bio-functionalized macro-porous Ti and the results showed a lower tendency to corrosion under sliding for bio-functionalized porous structures due to the load-carrying effect produced by the hard protruded porous $TiO_2$ surfaces formed by the MAO process.

Valuable insights are reported in the literature for bio-functionalized macro-porous Ti but most of the studies revealed limited information. The main limitation is related to the porous area that is exposed to the electrolyte, especially when the MAO process is proceeding. The inner pores may not be reached during the surface treatment and therefore, the oxide layer formed at the surface might be different throughout the macro-porous structure [48,55]. Besides, a lot is yet to be explored since the overall and long-term responses are yet to be understood. A better understanding of the degradation mechanics under fatigue or fretting corrosion solicitations for long times is not known for bio-functionalized macro-porous Ti. The effect of proteins, bio-organism, or bone ingrowth inside of the pores should also be understood.

## 6.5 SUMMARY AND CONCLUDING REMARKS

The search for an optimum solution to solve the clinical problems associated with the implants led to the manipulation of the implant structures in order to obtain hierarchical porosity. Bio-functionalized macro-porous Ti was presented as a promising solution in order to overcome the main problems of implant failures: infections, aseptic loosening, bioinertness of the implant surface, and osteolysis. The combination of macro-porous Ti with a micro-porous bioactive Ti oxide layer as a hierarchical structure can have the adjusted Young's modulus to match the one of the bone by controlling the implant macro-porosity and increasing the surface area that may improve the osseointegration. Additionally, through surface modification with MAO, it is possible to improve bioactivity, corrosion, and tribocorrosion performance of the implant by creating a micro-porous oxide layer containing bioactive elements.

The biological, mechanical, and triboelectrochemical performances of such a combination of solutions, bio-functionalized macro-porous Ti structures, are on the way to be improved and optimized for future implant applications. However, further long-term studies are needed to justify these approaches to be a candidate to substitute the current Ti-based implant materials. In addition to all the other mechanical properties of implant materials, the design of the implant should optimize the load transfer to the bone and such phenomena should not affect the osseointegration process. After that, in vivo studies are necessary to validate such tailored materials to be used as implants.

## REFERENCES

1. OECD, *Health at a Glance 2019*: OECD Indicators, Paris, 2019.
2. American Academy of Orthopaedic Surgeons, Total Hip Replacement, 2009.
3. L. Costa e Silva, M.I. Fragoso, J. Teles, Physical activity–Related injury profile in children and adolescents according to their age, maturation, and level of sports participation, *Sports Health.* 9 (2017) 118–125.
4. Statistical Office of the European Communities, Population Structure and Ageing - Statistics Explained, 2019, 1–10.
5. M.T. Mathew, C. Nagelli, R. Pourzal, A. Fischer, M.P. Laurent, J.J. Jacobs, M.A. Wimmer, Tribolayer formation in a metal-on-metal (MoM) hip joint: An electrochemical investigation, *J. Mech. Behav. Biomed. Mater.* 29 (2014) 199–212.
6. Y. Liao, R. Pourzal, M.A. Wimmer, J.J. Jacobs, A. Fischer, L.D. Marks, Graphitic tribological layers in metal-on-metal hip replacements, *Science (80-.).* 334 (2011) 1687–1690.
7. F.A. España, V.K. Balla, S. Bose, A. Bandyopadhyay, Design and fabrication of CoCrMo alloy based novel structures for load bearing implants using laser engineered net shaping, *Mater. Sci. Eng. C.* 30 (2010) 50–57.
8. H.J. Rack, J.I. Qazi, Titanium alloys for biomedical applications, *Mater. Sci. Eng. C.* 26 (2006) 1269–1277.
9. R.J. Ferguson, A.J. Palmer, A. Taylor, M.L. Porter, H. Malchau, S. Glyn-Jones, Hip and knee replacement 1- Hip replacement, *Lancet.* 392 (2018) 1662–1671.
10. S. Bilgen, G. Eken, Surgical site infection after total knee arthroplasty: A descriptive study, *Int. Multispecialty J. Heal.* 2 (2016) 1–8.
11. U. Holzwarth, G. Cotogno, Total hip arthroplasty: State of the art, challenges and prospects, *JRC Sci. Policy Rep.* (2012) Publications Office of the European Union, JRC72428.

12. G. Suzuki, S. Saito, T. Ishii, S. Motojima, Y. Tokuhashi, J. Ryu, Previous fracture surgery is a major risk factor of infection after total knee arthroplasty, knee surgery, *Sport. Traumatol. Arthrosc.* 19 (2011) 2040–2044.
13. M. Ribeiro, F.J. Monteiro, M.P. Ferraz, Infection of orthopedic implants with emphasis on bacterial adhesion process and techniques used in studying bacterial-material interactions, *Biomatter.* 2 (2012) 176–194.
14. M. Geetha, A.K. Singh, R. Asokamani, A.K. Gogia, Ti based biomaterials, the ultimate choice for orthopaedic implants – A review, *Prog. Mater. Sci.* 54 (2009) 397–425.
15. V. Goriainov, R. Cook, J. M. Latham, D. G. Dunlop, R.O.C. Oreffo, Bone and metal: An orthopaedic perspective on osseointegration of metals, *Acta Biomater.* 10 (2014) 4043–4057.
16. S.J. Macinnes, A. Gordon, J.M. Wilkinson, Risk factors for aseptic loosening following total hip arthroplasty. in *Recent Advances in Arthroplasty.* IntechOpen. 2012: 275–289.
17. B. Zhao, A.K. Gain, W. Ding, L. Zhang, X. Li, Y. Fu, A review on metallic porous materials: Pore formation, mechanical properties, and their applications, *Int. J. Adv. Manuf. Technol.* 95 (2018), 2641–2659.
18. S.D. Ulrich, T.M. Seyler, D. Bennett, R.E. Delanois, K.J. Saleh, I. Thongtrangan, M. Kuskowski, E.Y. Cheng, P.F. Sharkey, J. Parvizi, J.B. Stiehl, M.A. Mont, Total hip arthroplasties: What are the reasons for revision?, *Int. Orthop.* 32 (2008) 597–604.
19. A. Kumar, K. Biswas, B. Basu, Hydroxyapatite-titanium bulk composites for bone tissue engineering applications, *J Biomed Mater Res A.* 103 2015 791–806.
20. W. Simka, A. Iwaniak, G. Nawrat, A. Maciej, J. Michalska, K. Radwański, J. Gazdowicz, Modification of titanium oxide layer by calcium and phosphorus, *Electrochim. Acta.* 54 (2009) 6983–6988.
21. H. Guleryuz, H. Cimenoglu, Surface modification of a Ti-6Al-4V alloy by thermal oxidation, *Surf. Coatings Technol.* 192 (2005) 164–170.
22. E.A. Lewallen, S.M. Riester, C.A. Bonin, H.M. Kremers, A. Dudakovic, S. Kakar, R.C. Cohen, J.J. Westendorf, D.G. Lewallen, A.J. Wijnen, Biological strategies for improved osseointegration and osteoinduction of porous metal orthopedic implants, *Tissue Eng. Part B Rev.* 2 (2015) 218–230.
23. S.Y. Min, Y. Kim, T.K. Lee, S. Seo, H.S. Jung, D. Kim, J. Lee, B.J. Lee, The effect of porosity on the elasticity of pure titanium: An atomistic simulation, *Met. Mater. Int.* 16 (2010) 421–425.
24. O. Smorygo, A. Marukovich, V. Mikutski, A.A. Gokhale, G.J. Reddy, J.V. Kumar, High-porosity titanium foams by powder coated space holder compaction method, *Mater. Lett.* 83 (2012) 17–19.
25. A. Bansiddhi, D.C. Dunand, Titanium and NiTi Foams for Bone Replacement, in: Kajal Mallick (ed.), *Bone Substitute Biomaterials*, Woodhead Publishing Limited, Cambridge, England, 2014: pp. 142–179.
26. A.C. Alves, A.I. Costa, F. Toptan, J.L. Alves, I. Leonor, E. Ribeiro, R.L. Reis, A.M.P. Pinto, J.C.S. Fernandes, Effect of bio-functional MAO layers on the electrochemical behaviour of highly porous Ti, *Surf. Coat. Technol.* 386 (2020) 125487.
27. A.I. Costa, L. Sousa, A.C. Alves, F. Toptan, Tribocorrosion behaviour of bio-functionalized porous Ti surfaces obtained by two-step anodic treatment, *Corros. Sci.* 166 (2020) 108467.
28. A.R. Ribeiro, F. Oliveira, L.C. Boldrini, P.E. Leite, P. Falagan-lotsch, A.B.R. Linhares, Zambuzzi, B. Fragneaud, A.P.C. Campos, C.P. Gouvêa, B.S. Archanjo, C.A. Achete, E. Marcantonio, L.A. Rocha, J.M. Granjeiro, Micro-arc oxidation as a tool to develop multifunctional calcium-rich surfaces for dental implant applications, *Mater. Sci. Eng. C.* 54 (2015) 196–206.
29. J. Parthasarathy, B. Starly, S. Raman, A design for the additive manufacture of functionally graded porous structures with tailored mechanical properties for biomedical applications, *J. Manuf. Process.* 13 (2011) 160–170.

30. C. Simoneau, P. Terriault, B. Jetté, M. Dumas, V. Brailovski, Development of a porous metallic femoral stem: Design, manufacturing, simulation and mechanical testing, *Mater. Des.* 114 (2016) 546–556.
31. L. Lin, H. Wang, M. Ni, Y. Rui, T.-Y. Cheng, C.-K. Cheng, X. Pan, G. Li, C. Lin, Enhanced osteointegration of medical titanium implant with surface modifications in micro/nanoscale structures, *J. Orthop. Transl.* 2 (2014) 35–42.
32. M. Bruschi, D. Steinmüller-Nethl, W. Goriwoda, M. Rasse, Composition and modifications of dental implant surfaces, *J. Oral Implant.* (2015) 1–14.
33. G. Li, H. Cao, W. Zhang, X. Ding, G. Yang, Y. Qiao, X. Liu, X. Jiang, Enhanced osseointegration of hierarchical micro/nanotopographic titanium fabricated by microarc oxidation and electrochemical treatment, *ACS Appl. Mater. Interfaces.* 8 (2016) 3840–3852.
34. X. Zhang, D. Williams, IV - Biocompatibility and immune responses to biomaterials, in: *Definitions of Biomaterials for the Twenty-First Century*, Elsevier, Cambridge, MA, United States 2019: pp. 55–101.
35. X. Zhang, D. Williams, II - Biomaterials and biomedical materials, in: *Definitions of Biomaterials for the Twenty-First Century*, Elsevier, Cambridge, MA, United States 2019: pp. 15–23.
36. C. Aguilar, C. Guerra, S. Lascano, D. Guzman, P.A. Rojas, M. Thirumurugan, L. Bejar, A. Medina, Synthesis and characterization of Ti-Ta-Nb-Mn foams, *Mater. Sci. Eng. C.* 58 (2016) 420–431.
37. T. Hanawa, Titanium-tissue interface reaction and its control with surface treatment, *Front. Bioeng. Biotechnol.* 7 (2019) 170.
38. D.F. Williams, Biocompatibility in clinical practice: Predictable and unpredictable outcomes, *Prog. Biomed. Eng.* 1 (2019) 013001.
39. M. Niinomi, M. Nakai, J. Hieda, Development of new metallic alloys for biomedical applications, *Acta Biomater.* 8 (2012) 3888–3903.
40. A. Nouri, Titanium foam scaffolds for dental applications, *Met. Foam Bone.* In Metallic Foam Bone, Elsevier-Woodhead Publishing (2017) 131–160.
41. Y. Kirmanidou, M. Sidira, M.E. Drosou, V. Bennani, A. Bakopoulou, A. Tsouknidas, N. Michailidis, K. Michalakis, New Ti-alloys and surface modifications to improve the mechanical properties and the biological response to orthopedic and dental implants: A review, *Biomed Res. Int.* (2016).
42. L. Bolzoni, E.M. Ruiz-Navas, E. Gordo, Processing of elemental titanium by powder metallurgy techniques, *Mater. Sci. Forum.* 765 (2013) 383–387.
43. A. Nouri, P.D. Hodgson, C. Wen, Biomimetic porous titanium scaffolds for orthopedic and dental applications, in: Amitava Mukherjee (ed.), *Biomimetics Learning from Nature*, InTech, London, UK, 2010: pp. 415–451.
44. O. Addison, A.J. Davenport, R.J. Newport, S. Kalra, M. Monir, Do 'passive' medical titanium surfaces deteriorate in service in the absence of wear ?, *J. R. Soc. Interface.* 9 (2012) 3161–3164.
45. D.R.N. Correa, P.A.B. Kuroda, C.R. Grandini, L.A. Rocha, F.G.M. Oliveira, A.C. Alves, F. Toptan, Tribocorrosion behavior of β-type Ti-15Zr-based alloys, *Mater. Lett.* 179 (2016) 118–121.
46. Y. Tsutsumi, M. Ashida, K. Nakahara, A. Serizawa, H. Doi, C.R. Grandini, L.A. Rocha, T. Hanawa, Micro arc oxidation of Ti-15Zr-7.5Mo alloy, *Mater. Trans.* 57 (2016) 2015–2019.
47. J.R.S. Martins Júnior, R.A. Nogueira, R.O. De Araújo, T.A.G. Donato, V.E. Arana-Chavez, A.P.R.A. Claro, J.C.S. Moraes, M.A.R. Buzalaf, C.R. Grandini, Preparation and characterization of Ti-15Mo alloy used as biomaterial, *Mater. Res.* 14 (2011) 107–112.
48. A.C. Alves, R. Thibeaux, F. Toptan, A.M.P. Pinto, P. Ponthiaux, B. David, Impact of bio-functionalization on NIH/3T3 adhesion, proliferation and osteogenic differentiation of MC3T3-E1 over highly porous titanium implant material, *J. Biomed. Mater. Res. Part B - Appl. Biomater.* 107 (2019) 73–85.

49. Y. Torres, S. Lascano, J. Bris, J. Pavón, J.A. Rodriguez, Development of porous titanium for biomedical applications: A comparison between loose sintering and space-holder techniques, *Mater. Sci. Eng. C.* 37 (2014) 148–155.
50. S. Muñoz, J. Pavón, J.A. Rodríguez-Ortiz, A. Civantos, J.P. Allain, Y. Torres, On the influence of space holder in the development of porous titanium implants: Mechanical, computational and biological evaluation, *Mater. Charact.* 108 (2015) 68–78.
51. C. Domínguez-Trujillo, E. Peón, E. Chicardi, H. Pérez, J.A. Rodríguez-Ortiz, J.J. Pavón, J. García-Couce, J.C. Galván, F. García-Moreno, Y. Torres, Sol-gel deposition of hydroxyapatite coatings on porous titanium for biomedical applications, *Surf. Coatings Technol.* 333 (2017) 158–162.
52. C. Domínguez-Trujillo, F. Ternero, J.A. Rodríguez-Ortiz, J.J. Pavón, I. Montealegre-Meléndez, C. Arévalo, F. García-Moreno, Y. Torres, Improvement of the balance between a reduced stress shielding and bone ingrowth by bioactive coatings onto porous titanium substrates, *Surf. Coatings Technol.* 338 (2018) 32–37.
53. F. Matassi, A. Botti, L. Sirleo, C. Carulli, M. Innocenti, Porous metal for orthopedics implants, *Clin. Cases Miner. Bone Metab.* 10 (2013) 111–115.
54. S. Arabnejad, R. Burnett Johnston, J.A. Pura, B. Singh, M. Tanzer, D. Pasini, High-strength porous biomaterials for bone replacement: A strategy to assess the interplay between cell morphology, mechanical properties, bone ingrowth and manufacturing constraints, *Acta Biomater.* 30 (2016) 345–356.
55. F. Toptan, A.C. Alves, A.M.P. Pinto, P. Ponthiaux, Tribocorrosion behavior of bio-functionalized highly porous titanium, *J. Mech. Behav. Biomed. Mater.* 69 (2017) 144–152.
56. L. Yin, J. Zhou, L. Gao, C. Zhao, J. Chen, X. Lu, J. Wang, J. Weng, B. Feng, Characterization and osteogenic activity of $SrTiO_3/TiO_2$ nanotube heterostructures on microporous titanium, *Surf. Coatings Technol.* 330 (2017) 121–130.
57. J.S. Ansari, T. Takahashi, H. Pandit, Uncemented hips: Current status, *Orthop. Trauma.* 32 (2018) 20–26.
58. Y. Li, W. Wang, J. Duan, M. Qi, A super-hydrophilic coating with a macro/micro/nano triple hierarchical structure on titanium by two-step micro-arc oxidation treatment for biomedical applications, *Surf. Coatings Technol.* 311 (2017) 1–9.
59. C. Xiang, Y. Zhang, Z. Li, H. Zhang, Y. Huang, H. Tang, Preparation and compressive behavior of porous titanium prepared by space holder sintering process, *Procedia Eng.* 27 (2012) 768–774.
60. E. Baril, L.P. Lefebvre, S.A. Hacking, Direct visualization and quantification of bone growth into porous titanium implants using micro computed tomography, *J. Mater. Sci. Mater. Med.* 22 (2011) 1321–1332.
61. K.-H. Kim, N. Ramaswamy, Electrochemical surface modification of titanium in dentistry, *Dent. Mater. J.* 28 (2009) 20–36.
62. L. Le Guéhennec, A. Soueidan, P. Layrolle, Y. Amouriq, Surface treatments of titanium dental implants for rapid osseointegration, *Dent. Mater.* 23 (2007) 844–854.
63. T. Yang, H. Shu, H. Chen, C. Chung, J. He, Interface between grown osteoblast and micro-arc oxidized bioactive layers, *Surf. Coat. Technol.* 259 (2014) 185–192.
64. D. Wang, G. He, Y. Tian, N. Ren, W. Liu, X. Zhang, Dual effects of acid etching on cell responses and mechanical properties of porous titanium with controllable open-porous structure, *J. Biomed. Mater. Res. Part B Appl. Biomater.* 108 (2020) 1–10.
65. B.Q. Li, C.Y. Wang, X. Lu, Effect of pore structure on the compressive property of porous Ti produced by powder metallurgy technique, *Mater. Des.* 50 (2013) 613–619.
66. D.P. Mondal, M. Patel, S. Das, A.K. Jha, H. Jain, G. Gupta, S.B. Arya, Titanium foam with coarser cell size and wide range of porosity using different types of evaporative space holders through powder metallurgy route, *Mater. Des.* 63 (2014) 89–99.

67. D.K. Pattanayak, A. Fukuda, T. Matsushita, M. Takemoto, S. Fujibayashi, K. Sasaki, N. Nishida, T. Nakamura, T. Kokubo, Bioactive Ti metal analogous to human cancellous bone: Fabrication by selective laser melting and chemical treatments, *Acta Biomater.* 7 (2011) 1398–1405.
68. B. Zhang, X. Pei, C. Zhou, Y. Fan, Q. Jiang, A. Ronca, U. D'Amora, Y. Chen, H. Li, Y. Sun, X. Zhang, The biomimetic design and 3D printing of customized mechanical properties porous Ti6Al4V scaffold for load-bearing bone reconstruction, *Mater. Des.* 152 (2018) 30–39.
69. G. Li, L. Wang, W. Pan, F. Yang, W. Jiang, X. Wu, X. Kong, K. Dai, Y. Hao, In vitro and in vivo study of additive manufactured porous Ti6Al4V scaffolds for repairing bone defects, *Sci. Rep.* 6 (2016) 34072.
70. S. Wang, R. Li, L. Dong, Z.-Y. Zhang, G. Liu, H. Liang, Y. Qin, J. Yu, Y. Li, Fabrication of bioactive 3D printed porous titanium implants with Sr ions-incorporated zeolite coatings for bone ingrowth, *J. Mater. Chem. B.* 6 (2018) 3254–3261.
71. Y. Chino, D.C. Dunand, Directionally freeze-cast titanium foam with aligned, elongated pores, *Acta Mater.* 56 (2008) 105–113.
72. H. Do Jung, S.W. Yook, T.S. Jang, Y. Li, H.E. Kim, Y.H. Koh, Dynamic freeze casting for the production of porous titanium (Ti) scaffolds, *Mater. Sci. Eng. C.* 33 (2013) 59–63.
73. A. Ibrahim, F. Zhang, E. Otterstein, E. Burkel, Processing of porous Ti and Ti5Mn foams by spark plasma sintering, *Mater. Des.* 32 (2011) 146–153.
74. Y. Sakamoto, S. Moriyama, M. Endo, Y. Kawakami, Mechanical property of porous titanium produced by spark plasma sintering, *Key Eng. Mater.* 385–387 (2008) 637–640.
75. N. Tuncer, M. Bram, A. Laptev, T. Beck, A. Moser, H.P. Buchkremer, Study of metal injection molding of highly porous titanium by physical modeling and direct experiments, *J. Mater. Process. Technol.* 214 (2014) 1352–1360.
76. L. jian Chen, T. Li, Y. min Li, H. He, Y. hua Hu, Porous titanium implants fabricated by metal injection molding, *Trans. Nonferrous Met. Soc. China (English Ed.)* 19 (2009) 1174–1179.
77. Y. Chen, D. Kent, M. Bermingham, A. Dehghan-Manshadi, G. Wang, C. Wen, M. Dargusch, Manufacturing of graded titanium scaffolds using a novel space holder technique, *Bioact. Mater.* 2 (2017) 248–252.
78. A. Mansourighasri, N. Muhamad, A.B. Sulong, Processing titanium foams using tapioca starch as a space holder, *J. Mater. Process. Technol.* 212 (2012) 83–89.
79. X. Wang, Z. Lu, L. Jia, F. Li, Preparation and properties of low cost porous titanium by using rice husk as hold space, *Prog. Nat. Sci. Mater. Int.* 27 (2017) 344–349.
80. N. Taniguchi, S. Fujibayashi, M. Takemoto, K. Sasaki, B. Otsuki, T. Nakamura, T. Matsushita, T. Kokubo, S. Matsuda, Effect of pore size on bone ingrowth into porous titanium implants fabricated by additive manufacturing: An in vivo experiment, *Mater. Sci. Eng. C.* 59 (2016) 690–701.
81. B. Gorny, T. Niendorf, J. Lackmann, M. Thoene, T. Troester, H.J. Maier, In situ characterization of the deformation and failure behavior of non-stochastic porous structures processed by selective laser melting, *Mater. Sci. Eng. A.* 528 (2011) 7962–7967.
82. S. Van Bael, Y.C. Chai, S. Truscello, M. Moesen, G. Kerckhofs, H. Van Oosterwyck, J.P. Kruth, J. Schrooten, The effect of pore geometry on the in vitro biological behavior of human periosteum-derived cells seeded on selective laser-melted Ti6Al4V bone scaffolds, *Acta Biomater.* 8 (2012) 2824–2834.
83. E. Baril, L.-P. Lefebvre, Y. Thomas, Interstitial elements in titanium powder metallurgy: Sources and control, *Powder Metall.* 54 (2011) 183–186.
84. S. Özbilen, D. Liebert, T. Beck, M. Bram, Fatigue behavior of highly porous titanium produced by powder metallurgy with temporary space holders, *Mater. Sci. Eng. C.* 60 (2016) 446–457.

85. L.-P. Lefebvre, E. Baril, Effect of oxygen concentration and distribution on the compression properties on titanium foams, *Adv. Eng. Mater.* 10 (2008) 868–876.
86. L.-P. Lefebvre, E. Baril, L. de Camaret, The effect of oxygen, nitrogen and carbon on the microstructure and compression properties of titanium foams, *J. Mater. Res.* 28 (2013) 2453–2460.
87. F. Li, J. Li, T. Huang, H. Kou, L. Zhou, Compression fatigue behavior and failure mechanism of porous titanium for biomedical applications, *J. Mech. Behav. Biomed. Mater.* 65 (2016) 814–823.
88. F. Li, J. Li, H. Kou, G. Xu, T. Li, L. Zhou, Anisotropic porous titanium with superior mechanical compatibility in the range of physiological strain rate for trabecular bone implant applications, *Mater. Lett.* 137 (2014) 424–427.
89. B. Lee, T. Lee, Y. Lee, D.J. Lee, J. Jeong, J. Yuh, S.H. Oh, H.S. Kim, C.S. Lee, Space-holder effect on designing pore structure and determining mechanical properties in porous titanium, *Mater. Des.* 57 (2014) 712–718.
90. T. Imwinkelried, Mechanical properties of open-pore titanium foam, *Journal of Biomedical Materials Research Part A.* 81 (2007) 964–970.
91. V. Amigó, L. Reig, D. Busquets, J.L. Ortiz, J.A. Calero, Analysis of bending strength of porous titanium processed by space holder method, *Powder Metall.* 54 (2011) 67–70.
92. A. Zargarian, M. Esfahanian, J. Kadkhodapour, S. Ziaei-Rad, Numerical simulation of the fatigue behavior of additive manufactured titanium porous lattice structures, *Mater. Sci. Eng. C.* 60 (2016) 339–347.
93. S. Amin Yavari, R. Wauthle, J. Van Der Stok, A.C. Riemslag, M. Janssen, M. Mulier, J.P. Kruth, J. Schrooten, H. Weinans, A.A. Zadpoor, Fatigue behavior of porous biomaterials manufactured using selective laser melting, *Mater. Sci. Eng. C.* 33 (2013) 4849–4858.
94. M. Radovic, E. Lara-Curzio, L. Riester, Comparison of different experimental techniques for determination of elastic properties of solids, *Mater. Sci. Eng. A.* 368 (2004) 56–70.
95. Y. Torres, J.A. Rodríguez, S. Arias, M. Echeverry, S. Robledo, V. Amigo, J., Processing, characterization and biological testing of porous titanium obtained by space-holder technique, *J. Mater. Sci.* 47 (2012) 6553–6564.
96. R. Zdero, Experimental methods in orthopaedic biomechanics, 24th October 2016, 2017.
97. S. Kashef, S.A. Asgari, P.D. Hodgson, W. Yan, Simulation of three-point bending test of titanium foam for biomedical applications, *Front. Mater. Sci. Technol.* 32 (2008) 237–240.
98. R.A. Zavanelli, G.E.P. Henriques, I. Ferreira, J.M.D. De Almeida Rollo, Corrosion-fatigue life of commercially pure titanium and Ti-6Al-4V alloys in different storage environments, *J. Prosthet. Dent.* 84 (2000) 274–279.
99. M. Morita, T. Sasada, H. Hayashi, Y. Tsukamoto, The corrosion fatigue properties of surgical implants in a living body, *J. Biomed. Mater. Res.* 22 (1988) 529–540.
100. R.A. Antunes, M.C.L. De Oliveira, Corrosion fatigue of biomedical metallic alloys: Mechanisms and mitigation, *Acta Biomater.* 8 (2012) 937–962.
101. C. Leinenbach, C. Fleck, D. Eifler, The cyclic deformation behaviour and fatigue induced damage of the implant alloy TiAl6Nb7 in simulated physiological media, *Int. J. Fatigue.* 26 (2004) 857–864.
102. C. Leinenbach, D. Eifler, Fatigue and cyclic deformation behaviour of surface-modified titanium alloys in simulated physiological media, *Biomaterials.* 27 (2006) 1200–1208.
103. R.A. Zavanelli, A.S. Guilherme, G.E. Pessanha-Henriques, M. Antônio De Arruda Nóbilo, M.F. Mesquita, Corrosion-fatigue of laser-repaired commercially pure titanium and Ti-6Al-4V alloy under different test environments, *J. Oral Rehabil.* 31 (2004) 1029–1034.

104. Y. Okazaki, S. Rao, Y. Ito, T. Tateishi, Corrosion resistance, mechanical properties, corrosion fatigue strength and cytocompatibility of new Ti alloys without Al and V, *Biomaterials.* 19 (1998) 1197–1215.
105. C. Fleck, D. Eifler, Corrosion, fatigue and corrosion fatigue behaviour of metal implant materials, especially titanium alloys, *Int. J. Fatigue.* 32 (2010) 929–935.
106. C.A.H. Laurindo, R.D. Torres, S.A. Mali, J.L. Gilbert, P. Soares, Incorporation of Ca and P on anodized titanium surface: Effect of high current density, *Mater. Sci. Eng. C.* 37 (2014) 223–231.
107. M.T. Mohammed, Z.A. Khan, A.N. Siddiquee, Surface modifications of titanium materials for developing corrosion behavior in human body environment: A review, *Procedia Mater. Sci.* 6 (2014) 1610–1618.
108. J.E.G. González, J.C. Mirza-Rosca, Study of the corrosion behavior of titanium and some of its alloys for biomedical and dental implant applications, *J. Electroanalytical Chem.* 471 (1999) 109–115.
109. S. Virtanen, I. Milošev, E. Gomez-Barrena, R. Trebše, J. Salo, Y.T. Konttinen, Special modes of corrosion under physiological and simulated physiological conditions, *Acta Biomater.* 4 (2008) 468–476.
110. J.L. Xu, X.F. Jin, J.M. Luo, Z.C. Zhong, Fabrication and properties of porous NiTi alloys by microwave sintering for biomedical applications, *Mater. Lett.* 124 (2014) 110–112.
111. R. Menini, M. Dion, S.K.V. So, M. Gauthier, L.-P. Lefebvre, Surface and corrosion electrochemical characterization of titanium foams for implant applications, *J. Electrochem. Soc.* 153 (2006) 13–21.
112. J. Fojt, L. Joska, J. Málek, Corrosion behaviour of porous Ti-39Nb alloy for biomedical applications, *Corros. Sci.* 71 (2013) 78–83.
113. A.C. Alves, I. Sendão, E. Ariza, F. Toptan, P. Ponthiaux, A.M.P. Pinto, Corrosion behaviour of porous Ti intended for biomedical applications, *J. Porous Mater.* 23 (2016) 1261–1268.
114. S. KT, L. ME, M. MT, Failure causes in total hip replacements: A review, *Austin J. Orthop. Rheumatol.* 5 (2018) 1064.
115. J. Villanueva, L. Trino, J. Thomas, D. Bijukumar, D. Royhman, M.M. Stack, M.T. Mathew, Corrosion, tribology, and tribocorrosion research in biomedical implants: Progressive trend in the published literature, *J. Bio- Tribo-Corrosion.* 3 (2017) 1–8.
116. J. Geringer, K. Kim, B. Boyer, Fretting corrosion in biomedical implants, in: Dieter Landolt and Stefano Mischler (eds.), *Tribocorrosion of Passive Metals and Coatings*, Woodhead Publishing Limited, Cambridge, England, 2011: pp. 401–423.
117. A. Revathi, A. Dalmau, A. Igual, C. Richard, G. Manivasagam, Degradation mechanisms and future challenges of titanium and its alloys for dental implant applications in oral environment, 76 (2017) 1354–1368.
118. F. Živić, N. Grujović, S. Mitrović, D. Adamović, V. Petrović, A. Radovanović, S. Đurić, N. Pali, Tribology in industry friction and adhesion in porous biomaterial structure, *Tribol. Ind.* 38 (2016) 361–370.
119. M. Gui, S. Bong, J. Moo, Influence of porosity on dry sliding wear behavior in spray deposited Al–6Cu–Mn/SiCp composite, *Mater. Sci. Eng. A.* 293 (2000) 146–156.
120. D.P. Mondal, S. Das, N. Jha, Dry sliding wear behaviour of aluminum syntactic foam, *Mater. Des.* 30 (2009) 2563–2568.
121. Y. Yan, A. Neville, D. Dowson, Biotribocorrosion—an appraisal of the time dependence of wear and corrosion interactions: II. Surface analysis, *J. Phys. D. Appl. Phys.* 39 (2006) 3206–3212.
122. N. Diomidis, S. Mischler, N.S. More, M. Roy, Tribo-electrochemical characterization of metallic biomaterials for total joint replacement, *Acta Biomater.* 8 (2012) 852–859.

123. M.J. Runa, M.T. Mathew, M.H. Fernandes, L.A. Rocha, First insight on the impact of an osteoblastic layer on the bio-tribocorrosion performance of Ti6Al4V hip implants, *Acta Biomater.* 12 (2015) 341–351.
124. S. Banerjee, K. Issa, B.H. Kapadia, R. Pivec, H.S. Khanuja, M.A. Mont, Systematic review on outcomes of acetabular revisions with highly-porous metals, *Int. Orthop.* 38 (2014) 689–702.
125. A.C. Alves, F. Oliveira, F. Wenger, P. Ponthiaux, J.-P. Celis, L.A. Rocha, Tribocorrosion behaviour of anodic treated titanium surfaces intended for dental implants, *J. Phys. D. Appl. Phys.* 46 (2013) 404001.
126. F.G. Oliveira, A.R. Ribeiro, G. Perez, B.S. Archanjo, C.P. Gouvea, J.R. Araújo, A.P.C. Campos, A. Kuznetsov, C.M. Almeida, M.M. Maru, C.A. Achete, P. Ponthiaux, J.-P. Celis, L.A. Rocha, Understanding growth mechanisms and tribocorrosion behaviour of porous TiO2 anodic films containing calcium, phosphorous and magnesium, *Appl. Surf. Sci.* 341 (2015) 1–12.
127. M.T. Mathew, P. Srinivasa Pai, R. Pourzal, A. Fischer, M.A. Wimmer, Significance of tribocorrosion in biomedical applications: Overview and current status, *Adv. Tribol.* (2009) 1687–5915.
128. B.C. Costa, A.C. Alves, F. Toptan, A.M. Pinto, L. Grenho, M.H. Fernandes, D.Y. Petrovykh, L.A. Rocha, P.N. Lisboa-Filho, Exposure effects of endotoxin-free titanium-based wear particles to human osteoblasts, *J. Mech. Behav. Biomed. Mater.* 95 (2019) 143–152.
129. M. Kulkarni, Y. Patil-Sen, I. Junkar, C. V Kulkarni, M. Lorenzetti, A. Iglič, Wettability studies of topologically distinct titanium surfaces, *Colloids Surf. B. Biointerfaces.* 129 (2015) 47–53.
130. Y. Huang, H. Qiao, X. Nian, X. Zhang, X. Zhang, G. Song, Z. Xu, H. Zhang, S. Han, Improving the bioactivity and corrosion resistance properties of electrodeposited hydroxyapatite coating by dual doping of bivalent strontium and manganese ion, *Surf. Coatings Technol.* 291 (2016) 205–215.
131. M. Xiao, Y.M. Chen, M.N. Biao, X.D. Zhang, B.C. Yang, Bio-functionalization of biomedical metals, *Mater. Sci. Eng. C.* 70 (2017) 1057–1070.
132. T. Hanawa, Research and development of metals for medical devices based on clinical needs, *Sci. Technol. Adv. Mater.* 13 (2012) 64102.
133. T. Hanawa, A comprehensive review of techniques for biofunctionalization of titanium, *J. Periodontal Implant Sci.* 41 (2011) 263–272.
134. N.C.M. Oliveira, C.C.G. Moura, D. Zanetta-Barbosa, D.B.S. Mendonça, L. Cooper, G. Mendonça, P. Dechichi, Effects of titanium surface anodization with CaP incorporation on human osteoblastic response, *Mater. Sci. Eng. C.* 33 (2013) 1958–1962.
135. J. Dai, J. Zhu, C. Chen, F. Weng, High temperature oxidation behavior and research status of modifications on improving high temperature oxidation resistance of titanium alloys and titanium aluminides: A review, *J. Alloys Compd.* 685 (2016) 784–798.
136. M. Wen, C. Wen, P. Hodgson, Y. Li, Improvement of the biomedical properties of titanium using SMAT and thermal oxidation, *Colloids Surf. B Biointerfaces.* 116 (2014) 658–665.
137. J.R. Birch, T.D. Burleigh, Oxides Formed on Titanium by Polishing, Etching, Anodizing, or Thermal Oxidizing, Corrosion. 56 (2000) 1233–1241.
138. L. Xu, K. Zhang, C. Wu, X. Lei, J. Ding, X. Shi, C. Liu, Micro-arc oxidation enhances the blood compatibility of ultrafine-grained pure titanium, *Materials (Basel).* 10 (2017) 1446.
139. Y. Wang, H. Yu, C. Chen, Z. Zhao, Review of the biocompatibility of micro-arc oxidation coated titanium alloys, *Mater. Des.* 85 (2015) 640–652.
140. I.S. Park, T.G. Woo, W.Y. Jeon, H.H. Park, M.H. Lee, T.S. Bae, K.W. Seol, Surface characteristics of titanium anodized in the four different types of electrolyte, *Electrochim. Acta.* 53 (2007) 863–870.

141. M.D. Roach, R.S. Williamson, I.P. Blakely, L.M. Didier, Tuning anatase and rutile phase ratios and nanoscale surface features by anodization processing onto titanium substrate surfaces, *Mater. Sci. Eng. C.* 58 (2016) 213–223.
142. D. Velten, V. Biehl, F. Aubertin, B. Valeske, W. Possart, J. Breme, Preparation of $TiO_2$ layers on cp-Ti and Ti6Al4V by thermal and anodic oxidation and by sol-gel coating techniques and their characterization, *J. Biomed. Mater. Res.* 59 (2002) 18–28.
143. M. Fazel, H.R. Salimijazi, M.A. Golozar, M.R. Garsivaz jazi, A comparison of corrosion, tribocorrosion and electrochemical impedance properties of pure Ti and Ti6Al4V alloy treated by micro-arc oxidation process, *Appl. Surf. Sci.* 324 (2015) 751–756.
144. D. Quintero, O. Galvis, J.A. Calderón, J.G. Castaño, F. Echeverría, Control of the physical properties of anodic coatings obtained by plasma electrolytic oxidation on Ti6Al4V alloy, *Surf. Coatings Technol.* 283 (2015) 210–222.
145. L. Benea, E. Mardare-Danaila, M. Mardare, J.-P. Celis, Preparation of titanium oxide and hydroxyapatite on Ti–6Al–4V alloy surface and electrochemical behaviour in bio-simulated fluid solution, *Corros. Sci.* 80 (2014) 331–338.
146. I. da S.V. Marques, M.F. Alfaro, N.C. da Cruz, M.F. Mesquita, C. Sukotjo, M.T. Mathew, V.A.R. Barão, Tribocorrosion behavior of biofunctional titanium oxide films produced by micro-arc oxidation: Synergism and mechanisms, *J. Mech. Behav. Biomed. Mater.* 60 (2016) 8–21.
147. L. Benea, E. Danaila, P. Ponthiaux, Effect of titania anodic formation and hydroxyapatite electrodeposition on electrochemical behaviour of Ti-6Al-4V alloy under fretting conditions for biomedical applications, *Corros. Sci.* 91 (2015) 262–271.
148. K.C. Kung, T.M. Lee, T.S. Lui, Bioactivity and corrosion properties of novel coatings containing strontium by micro-arc oxidation, *J. Alloys Compd.* 508 (2010) 384–390.
149. D.A. Torres-Cerón, F. Gordillo-Delgado, S.N. Moya-Betancourt, Effect of the voltage pulse frequency on the structure of $TiO_2$ coatings grown by plasma electrolytic oxidation Effect of the voltage pulse frequency on the structure of $TiO_2$ coatings grown by plasma electrolytic oxidation, *J. Phys. Conf. Ser. Pap.* 935 (2017) 012067.
150. H.-T. Chen, C.-J. Chung, T.-C. Yang, C.-H. Tang, J.-L. He, Microscopic observations of osteoblast growth on micro-arc oxidized β titanium, *Appl. Surf. Sci.* 266 (2013) 73–80.
151. H.P. Felgueiras, L. Castanheira, S. Changotade, F. Poirier, S. Oughlis, M. Henriques, C. Chakar, N. Naaman, R. Younes, V. Migonney, J.P. Celis, P. Ponthiaux, L. a. Rocha, D. Lutomski, Biotribocorrosion (tribo-electrochemical) characterization of anodized titanium biomaterial containing calcium and phosphorus before and after osteoblastic cell culture, *J. Biomed. Mater. Res. Part B Appl. Biomater.* 103 (2015) 661–669.
152. N. Ohtsu, D. Ishikawa, S. Komiya, K. Sakamoto, Effect of phosphorous incorporation on crystallinity, morphology, and photocatalytic activity of anodic oxide layer on titanium, *Thin Solid Films.* 556 (2014) 247–252.
153. S. Durdu, M. Usta, The tribological properties of bioceramic coatings produced on Ti6Al4V alloy by plasma electrolytic oxidation, *Ceram. Int.* 40 (2014) 3627–3635.
154. S.A. Alves, R. Bayón, A. Igartua, V. Saénz de Viteri, L.A. Rocha, Tribocorrosion behaviour of anodic titanium oxide films produced by plasma electrolytic oxidation for dental implants, *Lubr. Sci.* 26 (2013) 7–8.
155. K. Korkmaz, The effect of Micro-arc Oxidation treatment on the microstructure and properties of open cell Ti6Al4V alloy foams, *Surf. Coatings Technol.* 272 (2015) 72–78.
156. Y. Gu, L. Chen, W. Yue, P. Chen, F. Chen, C. Ning, Corrosion behavior and mechanism of MAO coated Ti6Al4V with a grain-fined surface layer, *J. Alloys Compd.* 664 (2016) 770–776.
157. E. Santos, G.B. de Souza, F.C. Serbena, H.L. Santos, G.G. de Lima, E.M. Szesz, C.M. Lepienski, N.K. Kuromoto, Effect of anodizing time on the mechanical properties of porous titania coatings formed by micro-arc oxidation, *Surf. Coatings Technol.* 309 (2017) 203–211.

158. R. Zhou, D. Wei, J. Cao, W. Feng, S. Cheng, Q. Du, B. Li, Y. Wang, D. Jia, Y. Zhou, Conformal coating containing Ca, P, Si and Na with double-level porous surface structure on titanium formed by a three-step microarc oxidation, *RSC Adv.* 5 (2015) 28908–28920.
159. Z.S. Tao, W.S. Zhou, X.W. He, W. Liu, B.L. Bai, Q. Zhou, Z.L. Huang, K.K. Tu, H. Li, T. Sun, Y.X. Lv, W. Cui, L. Yang, A comparative study of zinc, magnesium, strontium-incorporated hydroxyapatite-coated titanium implants for osseointegration of osteopenic rats, *Mater. Sci. Eng. C.* 62 (2016) 226–232.
160. J. Zhou, L. Zhao, Multifunction Sr, Co and F co-doped microporous coating on titanium of antibacterial, angiogenic and osteogenic activities, *Acta Biomater.* 43 (2016) 358–368.
161. W. Zhang, H. Cao, X. Zhang, G. Li, Q. Chang, J. Zhao, Y. Qiao, X. Ding, G. Yang, X. Liu, X. Jiang, A strontium-incorporated nanoporous titanium implant surface for rapid osseointegration, *Nanoscale.* 8 (2016) 5291–5301.
162. W. Liu, M. Cheng, T. Wahafu, Y. Zhao, H. Qin, J. Wang, X. Zhang, L. Wang, The in vitro and in vivo performance of a strontium-containing coating on the low-modulus Ti35Nb2Ta3Zr alloy formed by micro-arc oxidation, *J. Mater. Sci. Mater. Med.* 26 (2015) 203.
163. X. Lin, X. Yang, L. Tan, M. Li, X. Wang, Y. Zhang, K. Yang, Z. Hu, J. Qiu, In vitro degradation and biocompatibility of a strontium-containing micro-arc oxidation coating on the biodegradable ZK60 magnesium alloy, *Appl. Surf. Sci.* 288 (2014) 718–726.
164. C. Wolf-Brandstetter, S. Urbanek, R.B. Beutner, D. Scharnweber, C.M. Moseke, Integration of strontium and copper species into calcium phosphate coatings and their cell biological characterization, 2017.
165. S. Amin Yavari, L. Loozen, F.L. Paganelli, S. Bakhshandeh, K. Lietaert, J.A. Groot, A.C. Fluit, C.H.E. Boel, J. Alblas, H.C. Vogely, H. Weinans, A.A. Zadpoor, Antibacterial behavior of additively manufactured porous titanium with nanotubular surfaces releasing silver ions, *ACS Appl. Mater. Interfaces* 8 (2016) 27.
166. A.D. Anastasiou, M. Nerantzaki, E. Gounari, M.S. Duggal, P. V. Giannoudis, A. Jha, D. Bikiaris, Antibacterial properties and regenerative potential of Sr2+ and Ce3+ doped fluorapatites; a potential solution for peri-implantitis, *Sci. Rep.* 9 (2019) 1–11.
167. S. Bose, D. Banerjee, A. Shivaram, S. Tarafder, A. Bandyopadhyay, Calcium phosphate coated 3D printed porous titanium with nanoscale surface modification for orthopedic and dental applications, *Mater. Des.* 151 (2018) 102–112.
168. J. Zhou, X. Wang, L. Zhao, Antibacterial, angiogenic, and osteogenic activities of Ca, P, Co, F, and Sr compound doped titania coatings with different Sr content, *Sci. Rep.* 9 (2019) 1–11.
169. H. Ishizawa, M. Ogino, Formation and characterization of anodic titanium oxide films containing Ca and P, *J. Biomed. Mater. Res.* 29 (1995) 65–72.
170. I.A.J. van Hengel, N.E. Putra, M.W.A.M. Tierolf, M. Minneboo, A.C. Fluit, L.E. Fratila-Apachitei, I. Apachitei, A.A. Zadpoor, Biofunctionalization of selective laser melted porous titanium using silver and zinc nanoparticles to prevent infections by antibiotic-resistant bacteria, *Acta Biomater.* 107 (2020) 325–337.
171. K. Glenske, P. Donkiewicz, A. Köwitsch, N. Milosevic-oljaca, Applications of metals for bone regeneration, *M. Appl. Met. Bone Regen.* 12;19 (2018) 826.
172. S.A. Alves, A.L. Rossi, A.R. Ribeiro, F. Toptan, A.M. Pinto, T. Shokuhfar, J.P. Celis, L.A. Rocha, Improved tribocorrosion performance of bio-functionalized TiO2 nanotubes under two-cycle sliding actions in artificial saliva, *J. Mech. Behav. Biomed. Mater.* 80 (2018) 143–154.
173. Y. Yan, J. Sun, Y. Han, D. Li, K. Cui, Microstructure and bioactivity of Ca, P and Sr doped TiO2 coating formed on porous titanium by micro-arc oxidation, *Surf. Coatings Technol.* 205 (2010) 1702–1713.

174. X. Rao, J. Li, X. Feng, C. Chu, Bone-like apatite growth on controllable macroporous titanium scaffolds coated with microporous titania, *J. Mech. Behav. Biomed. Mater.* 77 (2017). 225–233.
175. J. Sun, Y. Han, K. Cui, Microstructure and apatite-forming ability of the MAO-treated porous titanium, *Surf. Coatings Technol.* 202 (2008) 4248–4256.
176. M. Khodaei, M. Meratian, O. Savabi, M. Fathi, H. Ghomi, The side effects of surface modification of porous titanium implant using hydrogen peroxide: Mechanical properties aspects, *Mater. Lett.* 178 (2016) 201–204.
177. S. Amin Yavari, S.M. Ahmadi, J. van der Stok, R. Wauthle, A.C. Riemslag, M. Janssen, J. Schrooten, H. Weinans, A.A. Zadpoor, Effects of bio-functionalizing surface treatments on the mechanical behavior of open porous titanium biomaterials, *J. Mech. Behav. Biomed. Mater.* 36 (2014) 109–119.
178. Z. Gorgin Karaji, R. Hedayati, B. Pouran, I. Apachitei, A.A. Zadpoor, Effects of plasma electrolytic oxidation process on the mechanical properties of additively manufactured porous biomaterials, *Mater. Sci. Eng. C.* 76 (2017) 406–416.

# 7 Snake Skin Tribology

## *Material, Surface Structure, Frictional Properties, and Biomimetic Implications*

*Stanislav N. Gorb and Elena V. Gorb*
Kiel University

**CONTENTS**



## 7.1 DIVERSITY OF SURFACE STRUCTURES OF BELLY SCALES IN SNAKES

The absence of extremities in snakes has an important tribological consequence for these animals, as their ventral body side is almost in continuous contact with the substrate. To enable undulating locomotion, the snake skin has to provide high friction in order to support the forward motion and simultaneously low enough friction to facilitate sliding along the substrate (Renous et al. 1985). In general, frictional properties of a contact pair are dependent on the (1) surface energy, (2) surface topography, and (3) material properties of both surfaces (Bowden and Tabor 1986, Scherge and Gorb 2001). Since only limited information on the surface energy of snake skin is available, hydrophobic properties are assumed based on the histochemistry and chemical analysis of the skin (Landmann 1979, Landmann et al. 1981, Lillywhite and Maderson 1982, Alibardi 2005, Lillywhite 2005). Previous studies reported a strong variety of microstructures on the skin surface of snakes (Leydig 1873, Picado 1931, Hoge and Santos 1953, Maderson 1964, 1965, 1972, Price 1982, Renous et al. 1985,

DOI: 10.1201/9781003096443-7

Bea and Fontarnau 1986, Irish et al. 1988, Chiasson and Lowe 1989, Price and Kelly 1989, Hazel et al. 1999, Arnold 2002, Gower 2003, Alibardi 2005, Berthé et al. 2009, Abdel-Aal et al. 2012, Schmidt and Gorb 2012), and the majority of previous authors suggested that the ventral scales in combination with their surface microstructure are of high relevance for the snake locomotion (Figure 7.1).

In order to test a possible correlation between the scale microstructure and preferred habitat, Schmidt and Gorb (2012) studied ventral scale surfaces in 41 snake species from three different families (Pythonidae, Boidae, and Elapidae). The species used in this study are adapted to four different habitats (terrestrial, fossorial, arboreal, and aquatic). To reveal possible functional adaptations of the skin at various body sites, dorsal scales have been additionally examined. According to the literature, dorsal scales often evolved interesting coloration and photonic effects rather than specific tribological effects (Spinner et al. 2013b). Schmidt and Gorb (2012) also provided a comprehensive literature review on the dorsal and ventral microstructure of snakes and other representatives of Lepidosauria. They have clearly demonstrated microstructure variations within a single species just at the level of a single scale and between dorsal and ventral scales (Figure 7.2). Interestingly, ventral scales show pronounced microstructure similarities at the family level (Schmidt and Gorb 2012). Many traits of ventral microstructures are highly conservative from an evolutionary point of view: they have been reported from representatives of various families. Most probably, they did not evolve convergently, but rather inherited from the basal pattern of snakes. Surprisingly, a snake-like ventral microstructure was revealed even in legless geckos *Lialis* that employ a slithering kind of locomotion (Spinner et al. 2013a). This fact may support the idea about the strong functional significance of this kind of microstructure.

The microstructure correlates to some extent with animal habitat (Schmidt and Gorb 2012) and with the preferred mode of locomotion (Rieser et al. 2021) (Figure 7.2). The longest denticulations usually occur on ventral scales of arboreal species. This character presumably leads to the frictional increase in the lateral direction, which may be important for living in arboreal habitats. Fossorial species, burrowing in dry sandy substrates, show rather smooth ventral scale surfaces. This condition of the scale surface may reduce friction with granular abrasive substrates. Obligatory aquatic species possess a derived microstructure differing from non-aquatic species. Here, oberhautchen cells are less interdigitated, the longitudinal orientation of the microstructure is less expressed, and the microstructure is not as flat as is the case of non-aquatic species. These features possibly evolved due to the reduced or lost contact with the solid substrate (Schmidt and Gorb 2012). In general, one may conclude that the ventral scale microstructure correlates with phylogenetic relationships, and also with the species' preferred habitat and tribological requirements.

## 7.2 INNER ARCHITECTURE AND MECHANICAL PROPERTIES OF SNAKE SKIN

In spite of the fact that snakes from different lineages possess very different body sizes, scale dimension, body shape, and body mass (Mattison 2008), previous

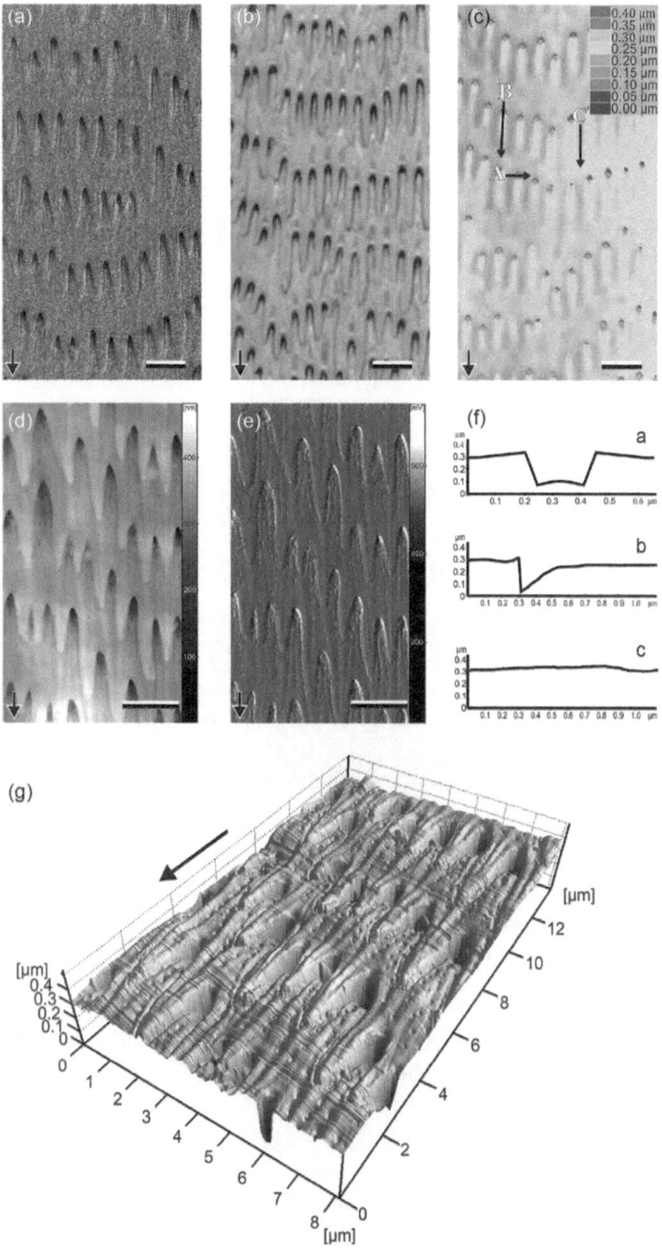


**FIGURE 7.1** Microstructure of a ventral scale in the snake *Lampropeltis getula californiae*. (a) Scanning electron microscopy (SEM) image. (b) Confocal laser scanning microscopy (CLSM) image, maximum intensity projection. (c) CLSM image, height map. (d) Atomic force microscopy (AFM) image, height image. (e) AFM image, error signal. (f) Three surface profiles based on AFM and CLSM data (for the sites on the skin, see c). (g) 3D image based on AFM data. The arrow points toward the posterior/caudal direction of the animal. Scale bars 2 μm. (From Baum et al. 2014a.)

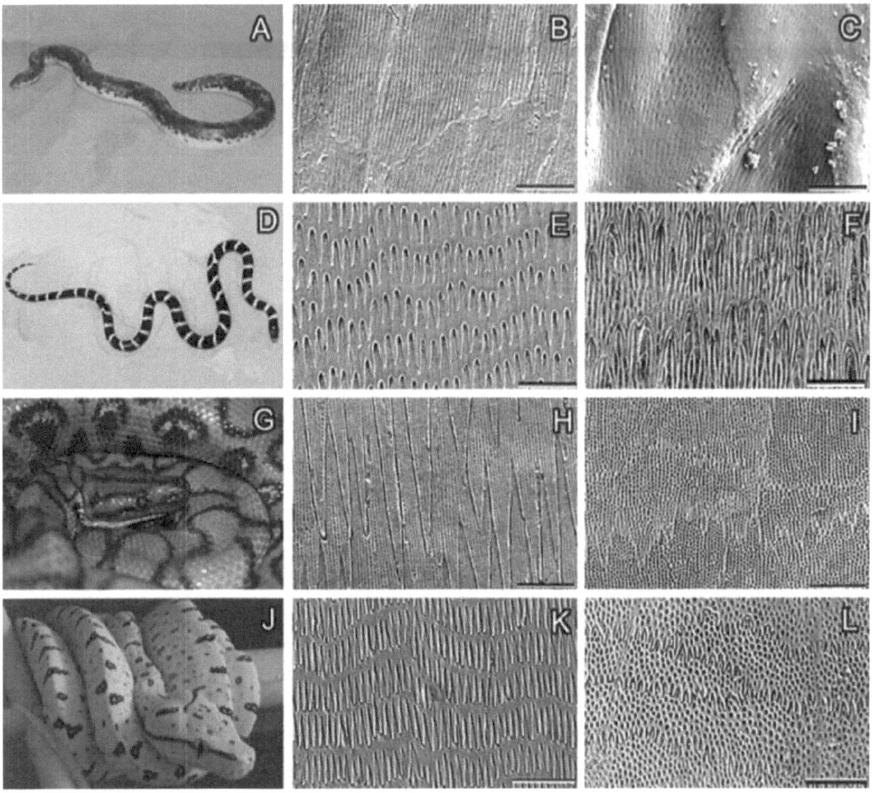


**FIGURE 7.2** Diversity of the snake skin microstructure. (a, d, g, j) Animals; (b, c, e, f, h, i, k, l) SEM images illustrating the surface microstructure of the ventral (b, e, h, k) and dorsal (c, f, i, l) scales. (a–c) Sand-burrowing *Gongylophis colubrinus*. (d–f) Terrestrial *Lampropeltis getula californiae*. (g–i) Generalist *Epicrates cenchria cenchria*. (j–l) Arboreal *Morelia viridis*. Scale bars 5 μm. (From Klein and Gorb 2012.)

morphological studies have clearly demonstrated that their epidermis consists of six layers overlying the dermis (Baden and Maderson 1970, Landmann 1979): (1) the oberhautchen layer, (2) β-layer, (3) mesos-layer, (4) α-layer, (5) lacunar tissue, and (6) clear layer (Figure 7.3). These structural data let previous authors assume that the snake epidermis consists of a hard, robust, inflexible outer surface (oberhautchen and β-layer) and a soft, flexible inner part (α-layers) (Baden and Maderson 1970, Landmann 1986, Mercer 1961, Maderson 1964, Alexander 1970, Matoltsy 1976, Wyld and Brush 1979, Fuchs and Marchuk 1983, Carver and Sawyer 1987, O'Guin et al. 1987, Alibardi and Toni 2006a–c). Nanoindentation experiments showed that both the effective elastic modulus and hardness of the outer and inner epidermis layers of the ventral scales of the Kenyan Sand Boa (*Gongylophis colubrinus*) exuvium are comparable with many other keratinous materials, such as horn, hoof, and nails (Klein et al. 2010). Additionally, the obtained results provided evidence for the presence of material properties gradient in the *G. colubrinus* skin (Klein et al. 2010).

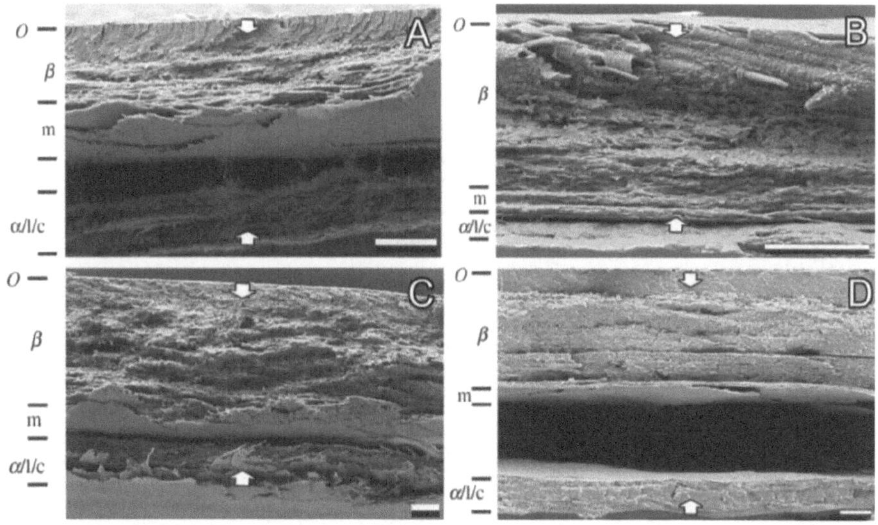


**FIGURE 7.3** SEM of cross fractures of the snake skin. The upper arrow in each image indicates the outer skin surface. The lower arrow in each image indicates the inner surface of the shed skin. O, Oberhäutchen; β, β-layer; m, mesos-layer; α, α-layer; l, lacunar tissue; c, clear layer. (a) *Gongylophis colubrinus.* (b) *Lampropeltis getula californiae.* (c) *Epicrates cenchria cenchria.* (d) *Morelia viridis.* Scale bars 5 μm. (From Klein and Gorb 2012.)

Why are these gradients important? In sliding contacts, local stress concentrations on the surface affect both material fatigue and failure. A surface is usually more effective against abrasive wear, if it has a gradient in material structure and properties from hard outer layers to more compliant inner layers, because such an architecture leads to a more uniform stress distribution in contact with substrate asperities, and thus to a minimization of local stress concentrations (Gibson and Ashby 1988, Wang and Weiner 1998, Suresh 2001, Bruet et al. 2008). A hard, uniform inflexible material easily forms cracks under pressure, whereas a uniform soft flexible material will be easily worn off under shear stress. It is assumed (Klein et al. 2010, Klein and Gorb 2012, 2016) (Figure 7.4) that the gradient in material properties from stiff surface layers to soft depth layers improves wear resistance in the snake skin by combining the advantages of stiffness/hardness and flexibility/compliance.

Although it is known that the epidermis of snakes consists of six main layers, there is only a single study comparing the distribution of these layers in the cross-sectional architecture of the skin between different snake species inhabiting different environments and preferably using different modes of locomotion (Klein and Gorb 2012). In a living snake, each scale has two exposed sides (called *the outer scale layer* (OSL)), which in turn can be subdivided into (1) *the outer scale surface* (OSS) that is in contact with the substrate and (2) *the inner scale surface* (ISS) that is in contact with the following scale. Both the OSS and ISS are composed of the six cell layers mentioned above. The non-exposed layer facing the epidermis side is called *the inner scale layer* (ISL), which becomes exposed after the skin has been shed. Klein and Gorb (2012) compared the inner structure and material properties

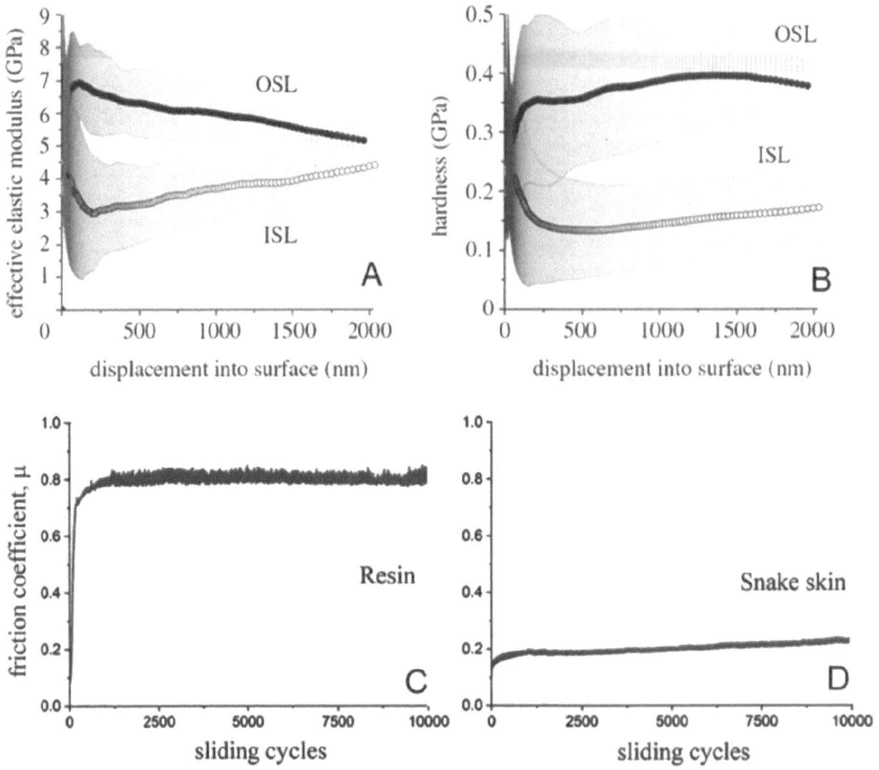


**FIGURE 7.4** (a and b). Results of nanoindentation measurements for both the outer (OSL) and inner (ISL) skin layers in the snake *Gongylophis colubrinus* (see the SEM image of the skin fracture in Figure 7.3a). (a) Effective elastic modulus versus displacement. (b) Hardness versus displacement. The error bars denote standard deviations. (c and d) Frictional coefficient curves for (c) the epoxy resin reference and (d) the ventral skin of *Lampropeltis getula californiae* snake at 1 N load during reciprocating sliding motion. (a, b. From Klein and Gorb 2012.) (c, d. From Sanchez-López et al. 2020.)

of the OSLs and ISLs of the exuvium epidermis of four snake species occupying different habitats: *Lampropeltis getula californiae* (terrestrial), *Epicrates cenchria cenchria* (generalist), *Morelia viridis* (arboreal), and *G. colubrinus* (sand-burrowing) (Figure 7.3). Scanning electron microscopy (SEM) of skin cross-sections revealed a strong variation in the epidermis structure between species. Interestingly, the difference between species in effective elastic modulus and hardness of the OSL and ISL measured by the nanoindentation was not large compared with the difference in epidermis thickness and architecture.

The friction and wear properties of the snake skin are indeed rather special. In a recent study, strong differences in friction and wear between snake skin and an epoxy resin having similar elasticity modulus have been demonstrated (Sanchez-Lopez et al. 2020). Snake skin showed considerably lower frictional coefficients stable over several thousands of sliding cycles and strong resistance to wear (Figure 7.4).

The wear mechanism in the resin reference, however, revealed rather severe wear. Additionally, the resin demonstrated strong stick-slip behavior, which was not notable on the snake skin. These results can be explained by three complementary mechanisms, such as the presence of (1) a fibrous layered composite material of the skin with graded material properties (Klein and Gorb 2012), (2) surface microstructure (Gray and Lissmann 1950, Hazel et al. 1999, Berthé et al. 2009, Schmidt and Gorb 2012, Baum et al. 2014a), and (3) ordered layers of lipid molecules at the surface (Baio et al. 2015) (see Section 7.3). The gradient in both elasticity modulus and hardness leads to protection of the surface against wear. The stress concentration in contact with abrasive particles of the high elastic modulus outer layers is reduced by the compliance of inner layers (Suresh 2001). The specific skin surface microstructure causes a reduction of contact area, which in turn leads to the reduction of adhesion and deformation in contact and thus to the reduction of frictional forces in the sliding direction (Berthé et al. 2009, Baum et al. 2014a). Furthermore, the lipid coating of the scale surface may provide both lubrication and wear protection of the ventral surface (Baio et al. 2015).

## 7.3 CHEMICAL ASPECTS OF THE SNAKE SKIN SURFACE

Detailed studies on the molecular structure of the snake epidermis have been previously carried out. X-ray diffraction has identified the presence of the two types of keratin (α- and β-layers) (Rudall 1974) and provided insight into the orientation of protein fibers within these epidermal regions (Baden et al. 1966). A later study focusing only on ventral scales of two species (*Notoechis scutatus* and *Bitis gabonica*) investigated the formation of keratins, associated β-proteins, and lipids within the keratin layers (Ripamonti et al. 2009). Three skin layers, differing in their content and molecular assembly of proteins and lipids, could be distinguished (Ripamonti et al. 2009). X-ray diffraction also suggested that the lipids in the mesos-layer are organized in a lamellar fashion and have aliphatic chains oriented perpendicular to the scale surface. The outer layers of the ventral scales contained both proteins with β-sheet conformation and lipids (Ripamonti et al. 2009). Immunocytochemical studies of the biological composition of skin of colubrid and viperid snake species revealed glycine-cysteine-rich corneous β-proteins accumulated in the oberhautchen cells. The amount of these proteins decreases in the underlying β- and α-layers (Alibardi 2014a–c).

Recently, the outermost surface of snake scales has been characterized using sum-frequency generation (SFG) spectra and near-edge X-ray absorption fine structure (NEXAFS) images collected from the shed skin of *L. g. californiae* epidermis (Baio et al. 2015). SFG's nonlinear optical selection rules provide information about the outermost surface, whereas NEXAFS probes the molecular structure of surfaces. This combined study has demonstrated that the β-layer (the oberhautchen), which is the outermost layer in contact with the environment, is covered with a nanometre thin lipid film with a specific structure differing in the dorsal and ventral scales.

Thus, the surface chemistries of the ventral and dorsal scales are unique according to their lipid coating (Baio et al. 2015). An ordered 'solid-like' lipid layer is present at the ventral scale versus a semi-disordered lipid layer on the dorsal surface

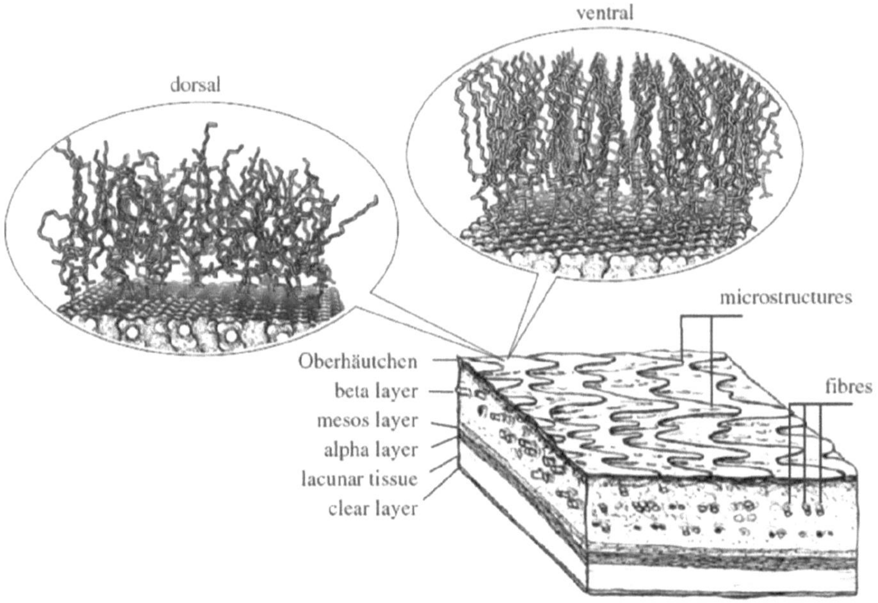


**FIGURE 7.5** Diagram of the lipid layer on the ventral and dorsal scale surfaces. The surface chemistries of the ventral and dorsal scales are unique. An ordered 'solid-like' lipid layer is present at the ventral scale versus a semi-disordered lipid layer on the dorsal surface. The terminology follows Klein and Gorb (2014). (From Baio et al. 2015.)

(Figure 7.5). Previous measurements of the frictional properties of dorsal and ventral scales from other species reported large differences between the ventral and dorsal scales (Gray and Lissmann 1950, Berthé et al. 2009, Hu et al. 2009) with ventral scales possessing a lower frictional coefficient than dorsal scales. This reduction of friction at the ventral surface could be a direct result of the observed well-ordered lipid monolayer acting as a lubricating layer. Previous studies of friction within vertebrate joints have revealed that adsorbed monolayers of synovial surfactants act as friction-reducing boundary lubricants (Hills and Butler 1984). In addition, lipids extracted from synovial fluid exhibit excellent anti-wear properties (Hills and Monds 1998, Scherge and Gorb 2001). This suggests that the ordered lipid layer observed at the ventral surface might have been optimized for both friction and wear reduction.

## 7.4 ANISOTROPIC FRICTION OF BELLY SCALES IN SNAKES: EXPERIMENTAL STUDIES

Experimental data on frictional properties of the snake skin have been obtained using a variety of approaches. Hazel et al. (1999) applied an atomic force microscope (AFM) for friction characterization at the level of a single microstructure. Gray and Lissmann (1950), Renous et al. (1985), Berthé et al. (2009), Hu et al. (2009), and Marvi and Hu (2012) used a sliding apparatus, whereas Abdel-Aal et al. (2012)

and Benz et al. (2012) employed a triboacoustic device and a microtribometer for friction measurements at the level of single and multiple scales, respectively.

In a recent series of publications, Baum et al. (2014a–c) studied the California King Snake (*L. g. californiae*) skin as a model system (Figure 7.1) and artificial surface microstructures inspired by snake skin. This species inhabits diverse habitats, such as forests, swamps, open grassland, as well as agricultural zones (Schmidt 2004). Since the snake is adapted to different environments, its surface microstructure might be specialized for a variety of substrates. The highly ordered denticulate microstructures of ventral scales are rather regular (Benz et al. 2012, Klein and Gorb 2012) (Figure 7.1). The skin of this species has a layered architecture, similar to that of other snakes (Landmann 1979) but contains highly ordered embedded fibers (Klein and Gorb 2014) (Figure 7.3b) and, as mentioned above (see Section 7.2), reveals a depth gradient in stiffness (Klein and Gorb 2012). Overall, the skin of this species is characterized only by little wear (Klein and Gorb 2012, 2014).

To investigate the role of the material stiffness gradient on the frictional coefficient and on an anisotropic character of the skin friction, dynamic friction measurements using a microtribometer on soft (cushioned) and stiff (uncushioned) epidermis have been carried out on various substrate roughness and in four different sliding directions (Baum et al. 2014a). Anisotropic frictional properties here are obviously not only caused by the surface topography but also supported by mechanical properties of the skin material itself, as previously shown for some technical surfaces (Bowden and Tabor 1986, Tramsen et al. 2018).

Frictional anisotropy along the longitudinal body axis (comparing frictional coefficient for forward and backward directions) and between longitudinal and transversal body axes was shown for *L. g. californiae* (Figure 7.6) (Baum et al. 2014a). The study by Berthé et al. (2009) did not reveal significant frictional anisotropy in forward and backward directions in *Corallus hortulanus*, whereas it reported significant differences in $\mu$ between the lateral and longitudinal directions of ventral scales.

As mentioned above, tribological data showed a strong influence of the effective elasticity modulus of the underlying material on anisotropic frictional properties of the snake skin in contact with smooth and rough surfaces. A possible explanation for the significant difference in frictional coefficients in lateral and cranial direction between hard- and soft-cushioned skin could be that the soft-cushioned samples are flexible enough to transfer the tangential stress applied in this direction into normal stress and vice versa. Furthermore, the energy of nanoimpacts at substrate irregularities during relative surface motion is presumably dissipated more effectively in the tribosystem with soft cushioning.

The observed frictional anisotropy is not only due to the asymmetric surface microstructure, but also presumably due to the remarkable ultrastructural architecture of the skin material, as it was recently shown for *L. g. californiae* (Klein and Gorb 2014). In longitudinal fractures of the skin, fibers oriented in caudal direction at an angle of about 15° to the ventral body surface were found (Figure 7.7). It is known that the fiber orientation is one of the most important variables that may explain reduced wear and optimized frictional behavior of composite materials (Lancaster 1968, Bhushan 2002). It is known from engineering materials that the frictional coefficient and wear are much smaller, if the fibers are oriented perpendicular to the

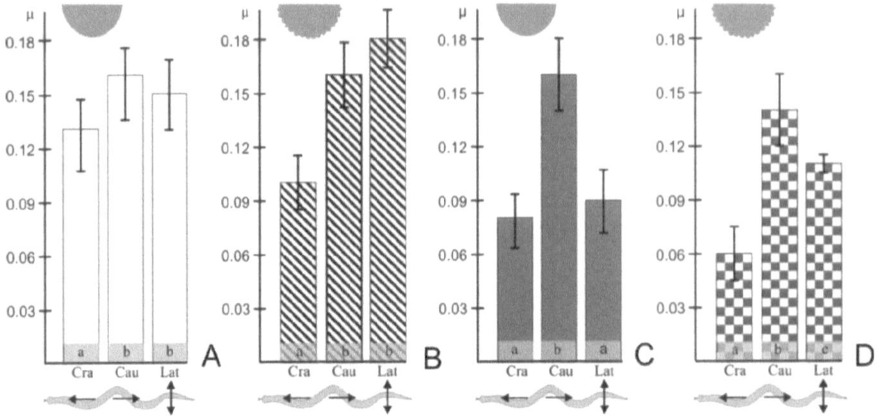


**FIGURE 7.6** Results of tribological measurements using smooth and rough glass balls. (a and c) Frictional coefficients between smooth glass ball and the uncushioned (a) and cushioned (c) exuvia samples. (b and d) Frictional coefficients between a rough glass ball and the uncushioned (b) and cushioned (d) exuvia samples. The anisotropy of the frictional properties was stronger in cushioned than in uncushioned samples. Semicircles indicate measurements with smooth or rough glass balls. Average frictional coefficients ($\mu$) and standard deviations are shown. The significantly different groups are marked by different letters (a–c). Arrow directions indicate how the experiment corresponds to the relative body movement of the snake. Cra, cranial direction; Cau, caudal direction; Lat, lateral direction. (From Baum et al. 2014a.)

sliding surface than if they are oriented parallel to the surface (Lancaster 1968). A combination of the asymmetric geometry of the surface microstructure and the sloped fiber orientation within the epidermis architecture could lead to anisotropic material properties, which could be an additional mechanism of the observed anisotropic frictional properties of the skin in some snakes.

It has been also shown that the substrate roughness has a strong influence on absolute values of the frictional coefficient. Independently on the type of sample cushioning, the frictional coefficient in the cranial direction is low in contact between the skin and rough glass ball. In the lateral direction, however, the snake skin in contact with the rough glass ball demonstrated significantly higher friction on hard-cushioned samples in comparison with soft-cushioned ones (Baum et al. 2014a). The substrate roughness is of high importance for the snake skin friction, because most natural surfaces are rough at the microscale and all surfaces are rough at the nanoscale. Roughness is the parameter determining whether the frictional behavior is dominated by molecular interactions, such as van der Waals' forces, or by mechanical interlocking of the two contact partners (Persson 2000, 2001, Persson and Volokitin 2006).

Berthé et al. (2009) have performed frictional measurements on scales of the Amazon Tree Boa *Corallus hortulanus* from different body parts in three directions on nine varying rough surfaces. They showed that the intermediate range of roughness ($R_a = 2.26$ μm) always led to the lowest frictional coefficient independently

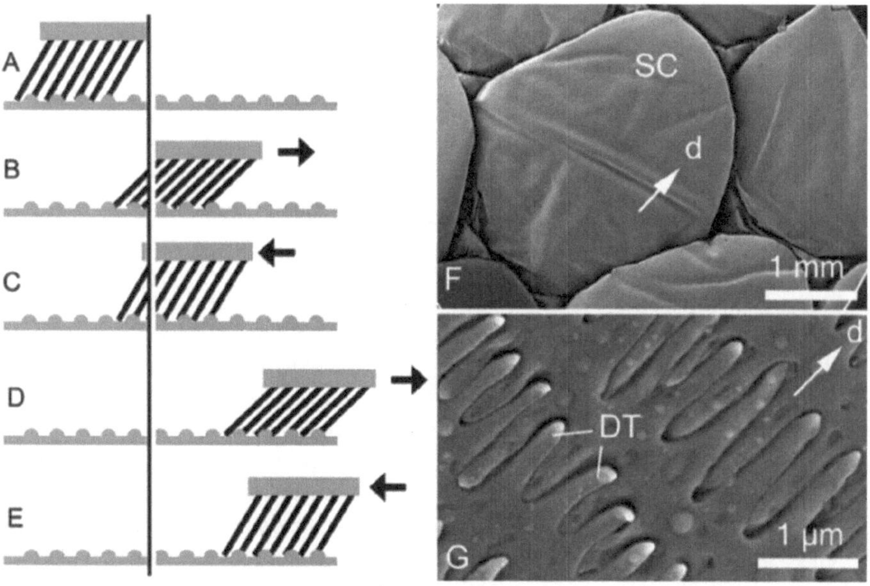


**FIGURE 7.7** Anisotropic system that assists body propulsion for locomotion. (a–e) Diagrams showing how a soft-embedded sloped stiff array of protuberances can generate propulsion along a non-smooth substrate. (f and g) Lateral scales of the snake *Python regius* at different magnifications in SEM. d, direction toward the tail (caudal); DT, denticulations; SC, scales. (Adapted from Filippov and Gorb 2013.)

on the specific site (ventral, lateral, and dorsal) on the snake skin tested. On eight from nine different substrate roughnesses, the anisotropy between backward direction (caudad) and lateral (sideward) was observed. It is necessary to note that the microstructure of *C. hortulanus* is totally different from that of *L. g. californiae.* In *C. hortulanus*, it is dominated by ridges along the body axis, which are separated by 300-nm wide grooves running more or less parallel to each other (Berthé et al. 2009). This surface microstructure is similar to that of *E .c. cenchria* (Figure 7.2H), whereas the microstructure of *L. g. californiae* is denticle-like (Figures 7.1 and 7.2e).

In another experiment, the frictional behavior of snake skin in contact with rigid styrofoam was tested by Marvi and Hu (2012) using a sliding apparatus. These authors also revealed anisotropic frictional properties in forward and backward directions of conscious and unconscious snakes. They did not investigate the role of microstructures in the context of friction control of snakes but noticed the tremendous ability of snakes to actively control frictional properties of their body surface. The authors have shown that the examined species, *Pantherophis guttatus* (Corn snake), is able to double its frictional coefficient on styrofoam just by active motion control of its scales.

Due to the fact that friction and wear cannot be considered as completely independent factors, it is rather probable that tribological optimization of the skin of *L. g. californiae* also bears some optimization to the wear reduction. Friction and wear mostly originate from two physical factors (Bowden and Tabor 1986): (1) the adhesion in regions of real contact (this force must be overcome to enable sliding) and (2) the plowing, deforming, or cracking of one of the surfaces. The relative influence of both factors depends on the contact pressure, material properties, and surface roughness of both surfaces in contact.

## 7.5 NUMERICAL MODELING OF THE FRICTIONAL-ANISOTROPY-BASED SYSTEM OF THE SNAKE SKIN

Surfaces covered with micro- and nanostructures, oriented at some angle to the plain, may cause strong mechanical anisotropy. Some of them also exhibit pronounced flexibility due to the material of the supporting layer or due to flexible connecting joints. Such structured biological systems have a wide range of functions including the transport of particles in insect cleaning devices and the propulsion generation during slithering locomotion of snakes. In this section, we report on numerical simulations studying the dependence of the anisotropic friction on (1) the slope of the structures, (2) rigidity of their joints, and (3) sliding speed. As reported in the previous sections, the snake skin consists of stiff asymmetrically oriented scales embedded in a flexible supporting layer (dermis). Additionally, on the surface of scales, there is a surface microstructure with a strongly asymmetric orientation, which provides additional frictional anisotropy of the skin on the substrate microasperities. The main function of such hierarchically organized anisotropic structures (scales and their surface microstructures) (Figure 7.7) is to reduce sliding friction in the forward sliding direction and to generate high propulsive force in the backward and sideward directions.

In the literature, there are only a few studies that report on frictional properties of snake skins on different roughnesses (Berthé et al. 2009, Hu et al. 2009, Abdel-Aal et al. 2012, Marvi and Hu 2012). Berthé et al. (2009) performed frictional experiments on different scales of *Corallus hortulanus* in three directions on nine different rough surfaces and showed frictional anisotropy along both the longitudinal and the transversal body axes on all tested substrate roughnesses. In another study, rough spheres with $R_a = 4$ μm (Abdel-Aal et al. 2012) and 2.4 μm (Baum et al. 2014) were used as a sliding probe. The frictional coefficient obtained in the cranial direction was always significantly lower than that obtained in the caudal direction. The snake skin sliding on a rigid styrofoam material (Marvi and Hu 2012) also showed anisotropic frictional properties along the longitudinal axis (forward and backward).

Snakes are also able to dynamically adapt their friction interactions by redistributing their local pressures and changing their winding angles, when either friction anisotropy is suppressed by the low-friction substrate or when an external force overcomes snake friction resistance on inclines. In order to understand these biotribology problems, a set of corresponding numerical models have been developed (Filippov and Gorb 2013, 2016).

It has been demonstrated that frictional anisotropy of the ventral surface of the snake skin only appears on substrates that have a characteristic range of roughness: either lower than or comparable with the dimensions of the skin microstructure (Filippov and Gorb 2016). This means that the scale microstructure should reflect an adaptation to the particular range of surface asperities of the substrate. For relatively

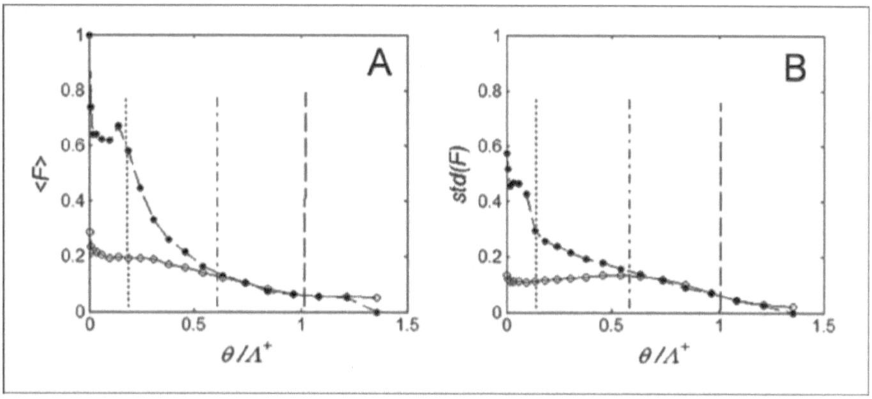


**FIGURE 7.8** Dependence of the mean frictional forces (a) and the standard deviations (b) for forward (white circles) and backward (black circles) directions of motion, calculated in the intervals between surface structure dimensions that correspond to the representative values. (From Filippov and Gorb 2016.)

small probes, friction is considerably higher for a positive direction of motion than in the opposite direction (Figure 7.8) (Filippov and Gorb 2016). It is interesting to note that a medium-sized probe leads to a practically complete smoothening of the friction curves in both directions. This means that frictional anisotropy only appears on the substrates that have a characteristic range of roughness, which is either lower than or comparable to the dimensions of the skin microstructure (Figure 7.8). However, in cases, where many substrate asperities interact simultaneously with the skin surface, the stick-slip behavior might not be as strongly pronounced as in the present model. It can be weaker due to the random distribution of denticle tips on the snake scales (Liley 1998). This mechanism can potentially reduce the stick-and-slip behavior of the snake skin.

In frictional experiments with anesthetized snakes on relatively smooth and rough surfaces ($R_a = 20$ and 200 µm, respectively), the rough surfaces demonstrated frictional anisotropy that almost completely disappeared on the smooth surfaces (Mühlberger et al. 2015). However, these experiments presumably showed the effect of the interlocking of individual scales on surfaces with coarse roughnesses. Thus, based on data from previous studies and the results of numerical modeling, we can assume that particular dimensions of the nanostructures on the ventral scales are adapted to enhance frictional anisotropy at microscale substrate roughnesses. Frictional anisotropy at the micro- and nanoscales is provided by the macro- and nanoscopic patterns on the ventral scales. It can therefore be concluded that the frictional anisotropy of the ventral surface is provided by two hierarchical levels of structures: scales and denticles. This is why snakes, whose locomotory ability is greatly decreased on smooth substrates, always rely on a certain dimension of roughness (even nanoscale roughness, where scales cannot be used) that might be sufficient for generating propulsion. This fact perfectly agrees with the results of numerical modeling.

## 7.6 SNAKE LOCOMOTION BASED ON FRICTIONAL-ANISOTROPY

In general, snakes possess four different modes of locomotion: (1) concertina, (2) rectilinear, (3) lateral undulation, and (4) sidewinding (Mattison 2008). Hu et al. (2009) visualized the dynamic load distribution of a snake during undulation and showed that the serpentine lateral undulation is characterized by propagating transverse waves along the snake's body from head to tail. Snakes are also able to dynamically change their frictional interactions with substrate surfaces by means of at least three different methods: (1) adjusting the angle of their scales (Marvi and Hu 2012), (2) redistributing their weight throughout various points of contact with the substrate (Hu et al. 2009; Marvi and Hu 2012), and (3) changing their winding angles (Alben 2013). Snakes can change their winding angles either when frictional anisotropy is suppressed by a particular roughness of the substrate or when the external force displacing the snake surpasses friction resistance during their locomotion on inclines. Numerical modeling was undertaken to understand this behavior and predict the snake's specific means of locomotion based on the interactions between the ventral surface of the snake skin and the substrate (Filippov et al. 2018). The adaptation of the winding curvature, considered in that model, represents an enhancement of friction anisotropy in critical behavioral situations, such as movement on low-friction substrates or while moving up and down a slope (Figure 7.9).

As stated above, in order to facilitate slithering or serpentine locomotion, snakes keep their ventral body surfaces in almost continuous contact with the substrate (Abdel-Aal 2018). The frictional forces generated by this contact are of crucial importance for propulsion generation. Due to their specific surface microstructure (Picado 1931, Hoge and Santos 1953, Maderson 1972, Irish et al. 1988, Chiasson and Lowe 1989, Price and Kelly 1989, Hazel et al. 1999, Gower 2003, Abdel-Aal et al. 2012, Schmidt and Gorb 2012), the ventral scales of snakes generate lower friction in the forward direction, which supports sliding, and higher friction in the lateral direction, enabling propulsive force generation during lateral winding (Renous et al. 1985, Berthé et al. 2009, Baum et al. 2014a–c). Since friction depends on the surface energy, material properties, and surface roughness of both bodies in contact (Bowden and Tabor 1986, Scherge and Gorb 2001), the slithering behavior of snakes should change on substrates with different surface properties.

It has been previously shown that at high speeds, snakes lift the curved parts of their bodies off the ground as they travel in lateral undulation and in sidewinding (Gans 1984, Jayne 1986). In sidewinding locomotion, an animal pushes into the direction, where the highest frictional coefficient applies. Recently, theoretical modeling has also predicted that snakes might be able to redistribute their weight and thereby concentrate their weight on specific points of contact (Hu et al. 2009). These points of contact approximately correspond to points of zero body curvature. Also, snakes are likely able to dynamically change their frictional interactions with a surface by adjusting the angles of their scales (Marvi and Hu 2012).

The friction anisotropy required for propulsion generation (Hu et al. 2009, Marvi and Hu 2012) may be rather sensitive to the roughness of the substrate, on which the snake moves, as mentioned above (Section 7.5) (Filippov and Gorb 2016). At specific relative dimensions of the snake skin microstructure and the substrate asperities,

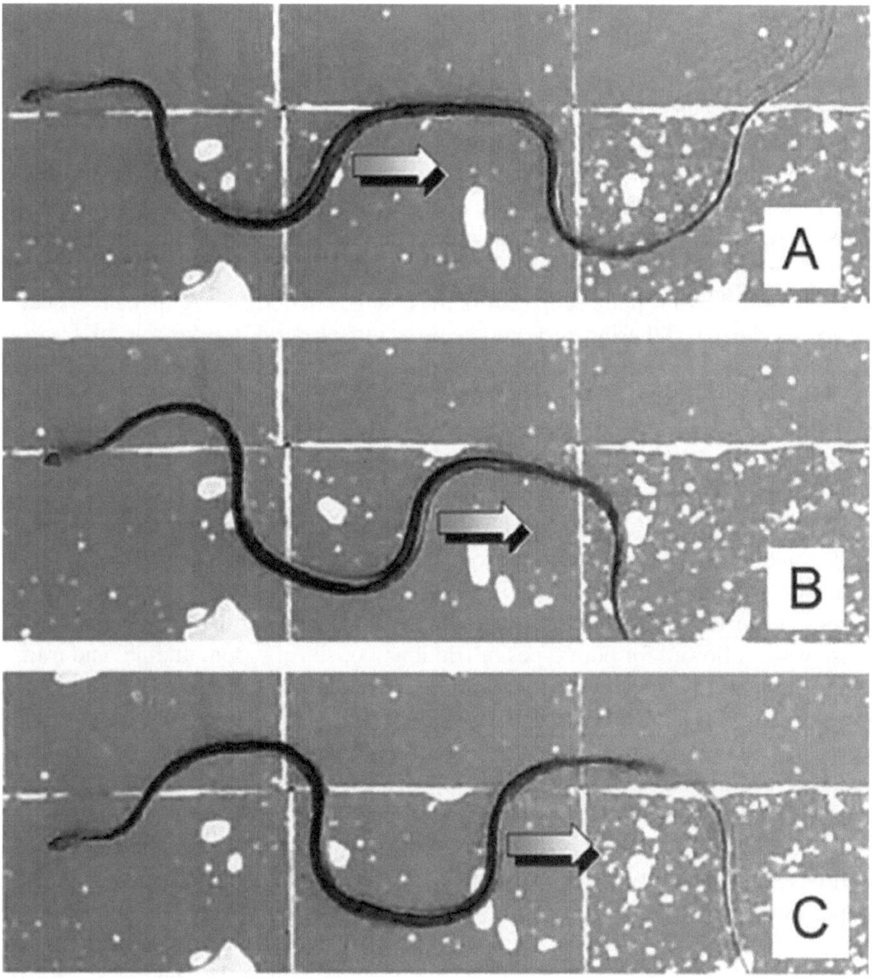


**FIGURE 7.9** (a–c) Specific configurations of the body shape of the snake *Psammophis schokari* during slithering locomotion on an inclined substrate with low friction. Note the typical shape of the snake body while moving on such a low-friction surface. It is moving almost without (or with minimal) forward propulsion. This can be easily recognized due to the snake's position relative to the markings on the substrate. Additionally, it is clear that typical soliton-like waves (arrow) propagate along the body against the direction of motion and new solitons at the cranial part of the body are generated as old ones at the caudal end of the body are annihilated. (From Filippov et al. 2018.)

friction anisotropy may be very low. The above model (Section 7.5) showed that the ventral surface of the snake skin demonstrates friction anisotropy only on the substrates with a characteristic range of roughness (either less than or comparable to the dimensions of the skin microstructure). This has an important tribological consequence for snake locomotion: at some substrate roughnesses, friction anisotropy may not support propulsion generation by normal slithering. In this case, a snake tries to adapt its body shape in a particular manner that enables the generation of propulsion (Figure 7.9). Recently, the behavioral adaptation of the snake to maintain friction anisotropy during locomotion on substrates with low friction or on inclines was numerically modeled (Filippov et al. 2018). Some previous snake locomotion models (Alben 2013, Wang et al. 2014) analyzed criteria of minimizing energy expenses during locomotion. This approach deepened our understanding of snake locomotion in general and is well suited for robotic applications. The approach by Filippov et al. (2018) is based on the analysis of specific locomotion patterns of real snakes based on the friction parameters of the ventral skin.

In the model by Filippov et al. (2018), the changes in body shape that a real snake undergoes during slithering locomotion were modeled taking the role of friction anisotropy into account. Previously in another numerical experiment, the general effect of the stiffness of surface structures on friction anisotropy (Filippov and Gorb 2013) was considered. In contrast to other snake locomotion models, where the snake body shape was presented as a sinusoidal or triangular wave (Wang et al. 2014) or some arbitrary smooth function minimizing the energy expenses functional (Alben 2013), the model by Filippov et al. (2018) considered the shape of a real snake during locomotion over substrates with low friction. The undulating method of snake locomotion was modeled by generating four solitary waves (two on each side), which correspond to the original action of the body with lost extremities (Figure 7.10). These waves allowed for both types of frictional anisotropy (longitudinal and transversal). Longitudinal friction anisotropy in snakes is limited due to the particular geometry of their skin microstructure, which allows for a maximum relationship between the longitudinal frictional forces in caudal/cranial directions of 1.75. This limitation requires an additional use of transversal anisotropy. In order to enhance transversal anisotropy, the snake changes its form factor (Alben 2013) (Figures 7.9 and 7.10). According to the model, however, the growth of transversal friction, which intuitively must magnify the propulsion during any undulating locomotion, enhances propulsion to only a limited extent. It was found that an increase of the ratio between transversal and longitudinal friction around 3 leads to weaker advantages in locomotion and higher energetic costs. These costs are related to the fact that the snake must continuously generate solitary waves in a caudal direction. The model showed that an increase of the wave amplitude makes the snake speed less dependent on the ratio between caudal and cranial friction, but entails higher energy costs (similar to the model of Wang et al. (2014)), and the wave amplitude increase did not even enhance the overall propulsion. This wave propagation pattern strongly differs from the motion of typical snake-inspired robots using wheels, because such robots do not create waves, but rather generate undulation by changing trajectory through turning their heads side-to-side.

## 7.7 BIOMIMETICS: FRICTION MICROSTRUCTURED SURFACES OF POLYMERS AND CERAMICS INSPIRED BY SNAKE SKIN

It is well known meanwhile that we can extend our creativity in engineering development by employing a huge bank of ideas from living nature. Every organism on

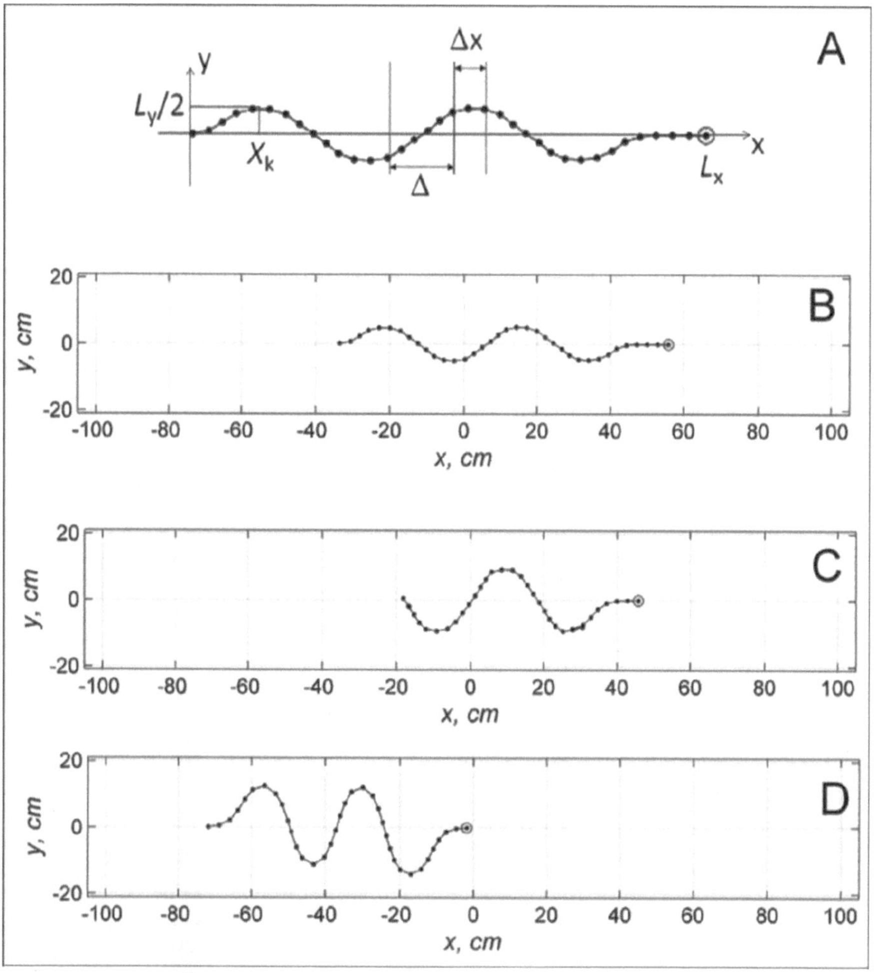


**FIGURE 7.10** Conceptual structure of the snake model. The snake segments are represented by black points. The "head" is indicated by the open circle. Some important shape parameters are shown in (a). Three different shapes of the snake corresponding to three different form factors (8.9, 3.7, 1.9) are presented in (b–d), respectively. (From Filippov et al. 2018.)

the Earth has evolved through adaptation and the survival of the fittest and, hence, organisms have retained only those evolutionary adaptations that make them better in their survival. Principles from snake skin tribology reported above might represent an interesting set of ideas for diverse engineering applications. Meanwhile, there are biomimetic surface structures with anisotropic friction inspired by the micro- (Greiner and Schäfer 2015) and nanostructures (Filippov and Gorb 2013, Baum et al. 2014c, Greiner and Schäfer 2015, Mühlberger et al. 2015) of the snake skin. For example, the so-called scale-like kirigami significantly enhances the crawling capability of a soft actuator. Recently reported highly stretchable anisotropic surfaces, in which mechanical instabilities induce a transformation from flat sheets to

3D-textured surfaces, similar to the scaled skin of snakes, is a much-promised engineering implementation of the snake scale principle (Rafsanjani et al. 2018).

A snake-inspired microstructured polymer surface (SIMPS) made of epoxy resin has been developed and characterized in contact with a smooth glass ball by a microtribometer in two perpendicular directions (Figure 7.11c–e). The SIMPS exhibited a considerable frictional anisotropy. Frictional coefficients measured along the microstructure were about 33% lower than those measured in the opposite direction (Baum et al. 2014c). The results demonstrate the existence of a common pattern of interaction between two general effects that influence friction: (1) molecular interaction depending on real contact area and (2) mechanical interlocking of both contacting surfaces. The strongest reduction of the frictional coefficient compared to the smooth reference surface was observed at a medium range of surface structure dimensions suggesting a trade-off between these two effects (Baum et al. 2014a–c).

Recently, a further step was successfully made in the replication of snake skin microstructure from a master onto a ceramic part via the fabrication of working stamps from snake skin (Mühlberger et al. 2015). Due to the size of the ceramic grains, the features of the 5 μm long denticles of the natural snake were difficult to

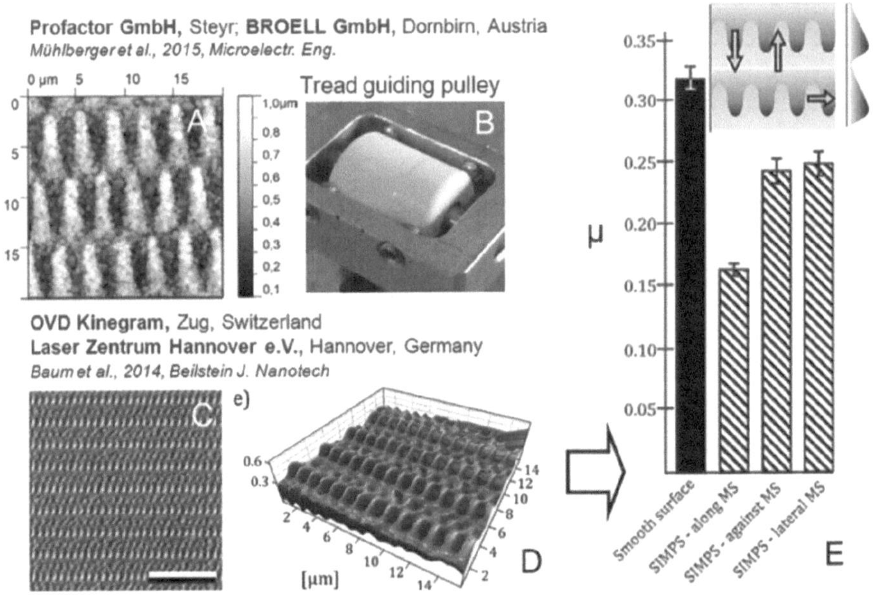


**FIGURE 7.11** Snake-skin-inspired engineering surfaces. (a and b). Anisotropic surface implemented into the ceramic thread guiding pulley in textile technology. (c and d). SEM (c) and AFM (d) micrographs of epoxy resin polymer molds of snake-inspired microstructured polymer surface (SIMPS). Scale bar 10 μm. (e) Results of tribological measurements of SIMPS in contact with a glass ball. Black column: smooth surface as a reference; hatched columns: tribological data on SIMPS. To investigate the influence of the anisotropic geometry of the microstructure on frictional properties, measurements were performed in three different sliding directions. MS, microstructure. Average frictional coefficients (μ) and standard deviations are shown. (a, b. From Mühlberger et al. 2015.) (c–e. From Baum et al. 2014c.)

replicate into the ceramics (Figure 7.11a and b). Therefore for biologically inspired artificial snake skin, the original biological structures were slightly enlarged and the applied process resulted in a successful replication of the asymmetric denticles into ceramic parts. The tribological testing showed strongly reduced friction in all tested conditions. This study has demonstrated that the transfer of complex biological structures onto the surfaces of non-flat ceramic parts is well possible. This opens great possibilities to employ patterns taken from functional biological surfaces, either directly or using bio-inspired shapes, into numerous real industrial applications.

## 7.8 FUTURE PERSPECTIVES

Additional research on the snake skin structure, surface chemistry, and mechanics of materials will help in the application of biological knowledge to recent tribological challenges in engineering. The incorporation of additional biological knowledge into the design of artificial systems will improve their performance. Unfortunately,

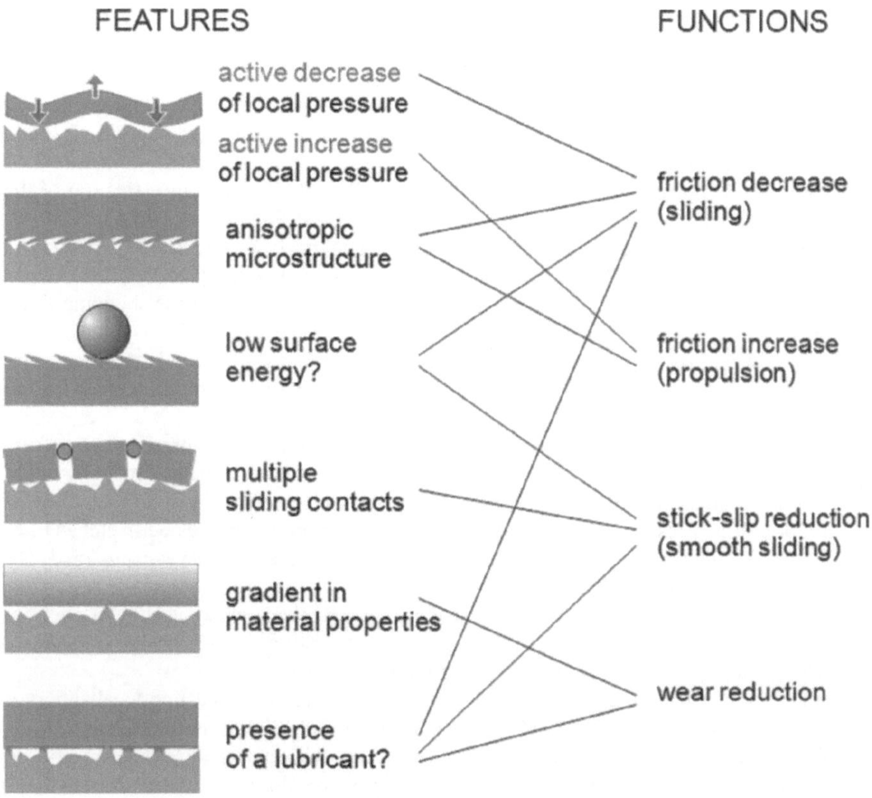


**FIGURE 7.12** Structure–function relationships of the snake skin (upper object) in contact with different substrates (lower object). Question marks (?) indicate features, which are not experimentally studied to date or which are under debate. Arrow showing up indicates active decrease of local pressure. Arrows showing down indicate active increase of local pressure.

biologists still do not have a complete understanding of the relationship between skin structure and its functional principles. We also do not know much about the exact snake skin performance *in vivo*. Hence, many technological areas will benefit from additional biological research. Additionally, a huge variety of snakes and their skins have never been comprehensively studied. Therefore, screening for new biological systems with interesting properties and their tribological adaptations to their environment (or to multiple environments) remains an extremely important research field in the nearest future.

Figure 7.12 shows some relationships between various features of the snake skin and their functional significance. There are some properties, which have been never studied before. For example, we do not know much about the surface energy of skins in different snakes, how they depend on a particular environment, and how they influence the tribological performance of the skin. How splitting of one big contact into multiple contacts with the substrate may potentially reduce stick-and-slip behavior during sliding of the snake skin? Very little is known about skin micropits, their particular distribution (Kovalev et al. 2016), and their tribological significance. Also, possible lubrication effects on the skin surface are not experimentally resolved. Finally, active control of local pressure needs further attention from experimental biologists.

## REFERENCES


Abdel-Aal, H. A. 2018. Surface structure and tribology of legless squamate reptiles. *J. Mech. Behavior Biomed. Mat.* 79:354–398.

Abdel-Aal, H. A., Vargiolu, R., Zahouani, H., et al. 2012. Preliminary investigation of the frictional response of reptilian shed skin. *Wear* 290–291:51–60.

Alben, S. 2013. Optimizing snake locomotion in the plane. *Proc. R. Soc. A.* 469:20130236.

Alexander, N. J. 1970. Comparison of α and β keratin in reptiles. *Z. Zellforsch.* 110:153–165.

Alibardi, L. 2005. Differentiation of snake epidermis, with emphasis on the shedding layer. *J. Morphol.* 264:178–190.

Alibardi, L. 2014a. Immunogold labeling shows that glycine–cysteine-rich beta-proteins are deposited in the Oberhäutchen layer of snake epidermis in preparation to shedding. *J. Morphol.* 276:144–151 (doi:10.1002/jmor.20327).

Alibardi, L. 2014b. Immunocytochemistry indicates that glycine-rich b-proteins are present in the beta-layer, while cysteine-rich beta-proteins are present in beta- and alpha-layers of snake epidermis. *Acta Zool.* 95:330–340 (doi:10.1111/azo.12030).

Alibardi, L. 2014c. Presence of a glycine-cysteine-rich beta-protein in the oberhautchen layer of snake epidermis marks the formation of the shedding layer. *Protoplasma* 251:1511–1520 (doi:10.1007/s00709-014-0655-7).

Alibardi, L., and Toni, M. 2006a. Cytochemical, biochemical and molecular aspects of the process of keratinization in the epidermis of reptilian scales. *Prog. Histochem. Cytochem.* 40:73–134 (doi:10.1016/j.proghi.2006.01.001).

Alibardi, L., and Toni, M. 2006b. Immunological characterization and fine localization of lizard beta-keratin. *J. Exp. Zool. B* 306:528–538 (doi:10.1002/jez.b.21105).

Alibardi, L., and Toni, M. 2006c. Immunolocalization and characterization of beta-keratins in growing epidermis of chelonians. *Tissue Cell* 38:53–63 (doi:10.1016/j.tice.2005.11.001).

Arnold, E. N. 2002. History and function of scale microornamentation in lacertid lizards. *J. Morphol.* 252:145–169.

Baden, H. P., and Maderson, P. F. 1970. Morphological and biophysical identification of fibrous proteins in the amniote epidermis. *J. Exp. Zool.* 174:225–232 (doi:10.1002/jez.1401740211).

Baden, H. P., Roth, S. I., and Bonar, L. C. 1966. Fibrous proteins of snake scale. *Nature* 212:498–499 (doi:10.1038/212498a0).

Baio, J., Spinner, M., Jaye, C., et al. 2015. Evidence of a molecular boundary lubricant at snakeskin surfaces. *J. R. Soc. Interface* 12:20150817.

Baum, M., Kovalev, A., Michels, J., et al. 2014a. Anisotropic friction of the ventral scales in the snake *Lampropeltis getula californiae. Tribol. Lett.* 54 (2):139–150. (doi:10.1007/s 11249-014-0319-y).

Baum, M., Heepe, L., and Gorb, S. N. 2014b. Friction behavior of a microstructured polymer surface inspired by snake skin. *Beilstein J. Nanotechnol.* 5:83–97.

Baum, M. J., Heepe, L., Fadeeva, E., et al. 2014c. Dry friction of microstructured polymer surfaces inspired by snake skin. *Beilstein J. Nanotechnol.* 5:1091–1103.

Bea, A., and Fontarnau, R. 1986. The study of the sloughing cycle in snakes by means of scanning electron microscopy. In *Studies in Herpetology,* ed. Z. Robek, pp. 373–376. Praque: Charles University.

Benz, M. J., Kovalev A. E., and Gorb, S. N. 2012. Anisotropic frictional properties in snakes. In *Bioinspiration, Biomimetics, and Bioreplication, Proc. of SPIE,* ed. A. Lakhtakia, and R. J. Martín-Palma, p. 83390X (doi:10.1117/12.916972).

Berthé, R. A., Westhoff, G., Bleckmann, H., et al. 2009. Surface structure and frictional properties of the Amazon tree boa *Corallus hortulanus* (Squamata, Boidae). *J. Comp. Physiol. A* 195:311–318.

Bhushan, B., 2002. Introduction to Tribology. New York: John Wiley & Sons.

Bowden, F.P., and Tabor, D. 1986. *The Friction and Lubrication of Solids.* Oxford: Clarendon Press.

Bruet, B. J. F., Song, J., Boyce, M. C., et al. 2008. Materials design principles of ancient fish armour. *Nat. Mat.* 7:1–9 (doi:10.1038/nmat2231).

Carver, W. E., and Sawyer, R. H. 1987. Development and keratinization of the epidermis in the common lizard, *Anolis carolinensis. J. Exp. Zool.* 243:435–443 (doi:10.1002/jez.1402430310).

Chiasson, R. B., and Lowe, C. H. 1989. Ultrastructural scale patterns in *Nerodia* and *Thamnophis. J. Herpetol.* 23:109–118.

Filippov, A. and Gorb,S.N. (2013) Frictional-anisotropy-based systems in biology: structural diversity and numerical model. Sci, Rep. 3(1240):1–6.

Filippov, A., and Gorb, S. N. 2016. Modelling of the frictional behaviour of the snake skin covered by anisotropic surface nanostructures. *Sci. Rep.* 6:23539.

Filippov, A. E., Westhoff, G., Kovalev, A., et al. 2018. Numerical model of the slithering snake locomotion based on the friction anisotropy of the ventral skin. *Tribol. Lett.* 66(3):119. (doi:10.1007/s11249-018-1072-4).

Fuchs, E., and Marchuk, D. 1983. Type I and type II keratins have evolved from lower eukaryotes to from the epidermal intermediate filaments in mammalian skin. *Proc. Natl Acad. Sci. USA* 80:5857–5861 (doi:10.1073/pnas.80.19.5857).

Gans, C. 1984. Slide-pushing: A transitional locomotor method of elongate squamates. *Symp. Zool. Soc. London* 52:12–26.

Gibson, L. J., and Ashby, M. F. 1988. *Cellular Solids: Structures and Properties.* New York: Pergamon Press.

Gower, D. J. 2003. Scale microornamentation of uropeltid snakes. *J. Morphol.* 258:249–268.

Gray, J., and Lissmann, H. W. 1950. The kinetics of locomotion of the grass-snake. *J. Exp. Biol.* 26:354–367.

Greiner, C., and Schäfer, M. 2015. Bio-inspired scale-like surface textures and their tribological properties. *Bioinspir. Biomim.* 10(4):044001 (doi:10.1088/1748-3190/10/4/044001).

Hazel, J., Stone, M., Grace, M. S., et al. 1999. Nanoscale design of snake skin for reptation locomotions via friction anisotropy. *J. Biomech.* 32:477–484.

Hills, B. A., and Butler, B. D. 1984. Surfactants identified in synovial fluid and their ability to act as boundary lubricants. *Ann. Rheumatic Dis.* 43:641–648 (doi:10.1136/ard.43.4.641).

Hills, B. A., and Monds, M. K. 1998. Enzymatic identification of the load-bearing boundary lubricant in the joint. *Br. J. Rheumatol.* 37:137–142.

Hoge, A. R., and Santos, P. S. 1953. Submicroscopic structure of "stratum corneum" of snakes. *Science* 118:410–411.

Hu, L. D., Nirody, J., Scott, T., et al. 2009. The mechanics of slithering locomotion. *PNAS* 106:10081–10085.

Irish, F. J., Williams, E. E., and Seling, E. 1988. Scanning electron microscopy of changes in epidermal structure occurring during the shedding cycle in squamate reptiles. *J. Morphol.* 197:105–126.

Jayne, B. C. 1986. Kinematics of terrestrial snake locomotion. *Copeia* 22:915–927.

Klein, M.-C.G., and Gorb, S. N. 2012. Epidermis architecture and material properties of the skin of four snake species. *J. R. Soc. Interface* 9(76):3140–3155.

Klein, M.-C. G., and Gorb, S. N. 2014. Ultrastructure and wear patterns of the ventral epidermis of four snake species (Squamata, Serpentes). *Zoology*117(5):295–314.

Klein, M.-C. G., and Gorb, S. N. 2016. Scratch resistance of the ventral skin surface in four snake species (Squamata, Serpentes). *Zoology* 119(2):81–96.

Klein, M.-C., Deuschle, J. K., and Gorb, S.N. 2010. Material properties of the skin of the Kenyan sand boa *Gongylophis colubrinus* (Squamata, Boidae). *J. Comp. Physiol. A* 196:659–668.

Kovalev, A., Filippov, A., and Gorb, S. N. 2016. Correlation analysis of symmetry breaking in the surface nanostructure ordering: case study of the ventral scale of the snake *Morelia viridis*. *Appl. Phys. A* 122(253):3–6.

Lancaster, J. K. 1968. The effect of carbon fibre reinforcement on the friction and wear of polymers. *J. Phys. D Appl. Phys.* 1:549–560.

Landmann, L. 1979. Keratin formation and barrier mechanisms in the epidermis of *Natrix natrix* (Reptilia, Sepentes): an ultrastructural study. *J. Morphol.* 162:93–126.

Landmann, L. 1986. Biology of the integument. In *The skin of Reptiles, Epidermis and Dermis*, ed. J. Bereither-Hahn, G. A. Matoltsy, and K. Sylvia-Richards, pp. 150–185. Heidelberg: Springer.

Landmann, L., Stolinski, C., and Martin, B. 1981. The permeability barrier in the epidermis of the grass snake during the resting stage of the sloughing cycle. *Cell Tissue Res.* 215:369–382.

Leydig, F. 1873. Über die äusseren Bedeckungen der Reptilien und Amphibien. *Archiv für mikroskopische Anatomie* 9(1):753–794.

Liley, M. 1998. Friction anisotropy and asymmetry of a compliant monolayer induced by a small molecular tilt. *Science* 280:273–275.

Lillywhite, H. B. 2005. Water relations of tetrapod integument. *J. Exp. Biol.* 209:202–226.

Lillywhite, H. B., and Maderson, P. F. A. 1982. Skin structure and permeability. In *Biology of the Reptilia*, ed. C. Gans, and F. H. Pough, pp. 379–442. New York: Academic Press.

Maderson, P. F. A. 1964. The skin of lizards and snakes. *Br. J. Herpetol.* 3:151–154.

Maderson, P. F. A. 1965. The structure and development of the squamate epidermis. In *The Biology of the Skin and Hair Growth*, ed. A. G. Lyne, and B. F. Short, pp. 129–153. Sydney: Angus and Robertson.

Maderson, P. F. A. 1972. When? why? and how?: some speculations on the evolution of the vertebrate integument. *Am. Zool.* 12:159–171.

Marvi, H., and Hu, D. L. 2012. Friction enhancement in concertina locomotion of snakes. *J. R. Soc. Interface* 9:3067–3080.

Matoltsy, A. G. 1976. Keratinization. *J. Invest. Dermatol.* 67:20–25 (doi:10.1111/1523-1747.ep12512473).

Mattison, C., 2008. Snakes. London: Dorling Kindersley Ltd.

Mercer, E. H. 1961. Keratin and keratinization. Oxford: Pergamon Press.

Mühlberger, M., Rohn, M., Danzberger, J., et al. 2015. UV-NIL fabricated bio-inspired inlays for injection molding to influence the friction behavior of ceramic surfaces. *Microelectronic Eng.* 141:140–144.

O'Guin, W. M., Galvin, S., Schermer, A., et al. 1987. Patterns of keratin expression define distinct pathways of epithelial development and differentiation. *Curr. Top. Dev. Biol.* 22:97–125 (doi:10.1016/S0070-2153(08)60100-3).

Persson, B. N. J. 2000. *Sliding Friction: Physical Principles and Applications.* Berlin: Springer.

Persson, B. N. J. 2001. Elastoplastic contact between randomly rough surfaces. *Phys. Rev. Lett.* 87:116101–116104.

Persson, B. N. J., and Volokitin, A. I. 2006. Rubber friction on smooth surfaces. *Euro. Phys. J. E* 21:69–80.

Picado, C. 1931. Epidermal microornaments of the Crotalinae. *Bull. Antivenin. Inst. Am.* 4:104–105.

Price, R. M. 1982. Dorsal snake scale microdermatoglyphics: ecological indicator or taxonomic tool? *J. Herpetol.* 16:294–306.

Price, R. M., and Kelly, P. 1989. Microdermatoglyphics: basal patterns and transition zones. *J. Herpetol.* 23:244–261.

Renous, S., Gasc, J. P., and Diop, A. 1985. Microstructure of the tegumentary surface of the Squamata (Reptilia) in relation to their spatial position and their locomotion. *Fortschr. Zool.* 30:487–489.

Ripamonti, A., Alibardi, L., Falini, G., et al. 2009. Keratin–lipid structural organization in the corneous layer of snake. *Biopolymers* 91:1172–1181 (doi:10.1002/bip.21184).

Rudall, K. M. 1974. X-ray studies of the distribution of protein chain types in the vertebrate epidermis. *Biochim. Biophys. Acta* 1:549–562 (doi:10.1016/0006-3002(47)90170-4).

Sánchez-López, J. C., Schaber, C. F., and Gorb, S. N. 2020. Long-term low friction maintenance and wear reduction on the ventral scales in snakes. *Mater. Lett.*:129011 (doi:10.1016/j.matlet.2020.129011).

Scherge, M., and Gorb, S. N. 2001. *Biological Micro- and Nanotribology.* Berlin: Springer.

Schmidt, D. 2004. *Die Kettennatter.* Lampropeltis getula. Münster: Natur und Tier-Verlag.

Schmidt, C. V., and Gorb, S. N. 2012. *Zoologica. Snake Scale Microstructure: Phylogenetic Significance and Functional Adaptations.* Stuttgart: Schweinsbart Science.

Spinner, M., Gorb, S. N., and Westhoff, G. 2013a. Diversity of functional microornamentation in slithering geckos *Lialis* (Pygopodidae). *Proc. R. Soc. B* 280:20132160.

Spinner, M., Kovalev, A., Gorb, S. N., et al. 2013b. Snake velvet black: hierarchical micro- and nanostructure enhances dark colouration in *Bitis rhinoceros. Sci. Rep.* 3:1846.

Suresh, S. 2001. Graded materials for resistance to contact deformation and damage. *Science* 292:2447–2451.

Rafsanjani, A., Zhang, Y. Liu, B., et al. 2018. Kirigami skins make a simple soft actuator crawl. *Sci. Robot.* 3(15):eaar7555. (doi:10.1126/scirobotics.aar7555).

Rieser, J. M., Li, T.-D., Tingle, J. L., et al. 2021. Functional consequences of convergently-evolved microscopic skin features on snake locomotion. *PNAS* 118(6), e2018264118.

Tramsen, H. T., Gorb, S. N., Zhang, H., et al. 2018. Inversion of friction anisotropy in a bio-inspired asymmetrically structured surface. *J. R. Soc. Interface* 15:20170629. (doi:10.1098/rsif.2017.0629).

Wang, R. Z., and Weiner, S. 1998. Strain–structure relations in human teeth using Moire fringes. *J. Biomech.* 31:135–141 (doi:10.1016/S0021-9290(97)00131-0).

Wang, X., Osborne, M. T., and Alben, S. 2014. Optimizing snake locomotion on an inclined plane. *Phys. Rev. E* 89:012717.

Wyld, J. A., and Brush, A. H. 1979. The molecular heterogeneity and diversity of reptilian keratins. *J. Mol. Evol.* 12:331–347 (doi:10.1007/BF01732028).

# 8 Surface Modification of Ti6Al4V through Electrical Discharge Machining Assisted Alloying to Improve Its Tribological Behavior—The Pathway to Genesis of a New Alloying Technique

*Jibin T Philip*
National Institute of Technology Mizoram
Amal Jyothi College of Engineering

*Basil Kuriachen*
National Institute of Technology Mizoram
National Institute of Technology Calicut

**CONTENTS**



DOI: 10.1201/9781003096443-8

## 8.1 INTRODUCTION

Titanium—a familiar name to the majority of the world mass, reigns as the ninth abundant material on Earth's crust. Its subtle innate material characteristics such as strength to weight ratio, biocompatibility, and resistance to corrosion specifically sets itself ahead of the majority of other advanced materials. The application base of this functional material is disparately multi-faceted, with a domineering upper hand in the fields viz. marine, aerospace, and biomedicine. Among the titanium alloy series, Ti6Al4V (Ti64) found its place as the heavily utilized material of the category due to its cutting-edge capabilities of heat treatability and strength, in particular. However, the material finds its implausible downfall in possessing elatedly poor tribo-properties. Moreover, the material displays the negate characteristic of thermal softening (TS) at elevated temperatures. Consequently, the range of definite applicability of the alloy has been delimitated to a finite range of scope. Removing such barriers by deriving valid solutions to such disparate situations can be a commendable act, amidst enormous efforts by countless researchers over the same domain. It can open doorways for the implementation and utilization of Ti64 in the discarded domains as well.

Surface modification techniques have been efficaciously applied to improve the friction and wear characteristics of Ti64, in the past. The methods include physical vapor deposition (PVD), chemical vapor deposition (CVD), nitriding, laser texturing, and so on. Nevertheless, the uneconomic and complex nature of the aforementioned mechanisms forces them to be sidelined from extensive applications. The basis for such failed conditioning points to the varying characteristic of Ti64 at diverse load, velocity, and environmental real-time situations. The dominant mechanisms that prevail to cause catastrophic effects on the material surface ensuing counter-body interactions are adhesion and abrasion-assisted delamination. Warding off such adverse conditions needs full-fledged understanding and insight into the underlying mechanisms instigating the material removal in Ti64 alloys. Thus, compatible surface refurbishing methods can be enforced to enfeeble the disintegrating tribo-characteristics of the material.

Electrical discharge machining (EDM) has embarked itself as one of the dominant methods in the processing of titanium alloys. The inherent characteristic of the process viz. capability in machining advanced and electrically conductive materials, potential in processing intricate shapes along with acceptable accuracy of the machined components, sets it apart from other non-conventional machining processes. The peculiar nature of EDM, often considered as a degrading factor to form a new alloyed layer on the substrate (generally referred to as the recast layer (RL)), is that it is very hard and non-etchable. The layer so formed possesses pores, microcracks, pockmarks, and cavities. The thickness of the RL is observed to vary largely with the process input parameters and conditions, viz. discharge current, discharge voltage, pulse on/off time, servo feed, dielectric fluid, electrode materials (tool/workpiece), and so on. The EDM machined surfaces are generally subjected to post-processing operations to remove the RL, on the consideration that the layer can have a degrading effect on the component performance during real-time implementation. Although it can be true for many of the materials, the efforts for positive exploitation of the thereby formed layer in the context of advanced materials are elatedly scarce in number. The exploration of the same for encapsulating the positive traces of the surface properties pertinent to the RL can prove phenomenal in disregarding any further essential surface modification requirements after EDM processing.

The extended investigations on the RL formed ensuing EDM machining has thrown light onto an in-depth understanding of the process as a whole. In general, the material removal in the EDM process takes place by the high velocity impinging electrons and ions striking the electrode surfaces. These electrons and ions activate the cavitation effect leading to localized material removal through melting and evaporation. The low flushing efficiency of the dielectrics often leads to the redeposition of a part of the removed material mixed with fused components from both the electrodes, dielectric, and the reaction byproducts. This RL can possess varying properties and capabilities depending upon the entrapped components in the re-solidified mass. So, the surface properties of the functional material can be controlled effectively by monitoring the process characteristics, materials (tool/workpiece electrodes), and environment. Furthermore, it displayed high toughness and yield strength in comparison to the unmachined Ti64 surfaces.

Given the above, the essentiality associated with the need for improving the tribological properties of Ti6Al4V is evident. Exploring the potential possibilities openly available for enhancing such specific material responses can expand the utility domain of the alloy. Besides, Ti6Al4V being a workhorse material imparting superior interactive characteristics can aid in implementing the alloy for dynamic wear-resistant applications. The potential of EDM to act as a combined machining and surface modification technique is still in the exploratory stage and requires precise attention to uncover its associated complexities. Reportedly, the RL developed through EDM has unique mechanical, metallurgical, and tribological properties, which can presumably aid in assisting the alloy to exhibit competent properties.

Hence, this work explores the tribo-behavior of Ti6Al4V (influential factors and interactive responses) and the ability of the EDM process to develop protective surface layers over the substrate.

## 8.2 TRIBOLOGY OF TI6AL4V

Ti6Al4V (Ti64) has broad applicability in the aerospace industry, chemical industry, automotive industry, manufacturing of ships and machinery, fabrication of parts/equipment for the food industry, oil/gas industry, and civil and biomedical engineering. The competent indigenous properties of the alloy include high strength to density ratio, superior corrosion resistance, absolute inertness to in-vivo conditions, and capacity to be knit with bones and tissues. Nonetheless, the machining of the Ti64 alloy using conventional techniques is arduous due to the reasons, viz. (1) low tool life caused by the high heat concentration at the cutting edge, since the alloy possesses low thermal conductivity (60%) less than commercially pure titanium (CP Ti), and (2) the strong chemical affinity with the cutting tool materials, leading to smearing and welding effect (particularly at high temperatures) and low modulus of elasticity. Besides, the Ti64 alloy is associated with inconsistent tribological characteristics due to its susceptibility to plastic shearing, low work/strain hardenability, and weak protection offered by the developed oxides during interactive sliding with distinct counter-faces. Hence, the alloy often gets neglected from being incorporated in dynamic systems demanding resistance to wear under interactive conditions.

### 8.2.1 Factors Affecting Tribo-Behavior

The tribo-behavior of the Ti64 alloy is governed by several internal and external factors related to the material. A few of those are discussed below.

#### 8.2.1.1 Thermal Oxidation (ThO)

Thermal oxidation (ThO) is a simple and inexpensive processing method used on material surfaces to form a relatively hard and dense oxide film [1]. The surface alteration of Ti64 was found to strengthen the alloy's tribo-behavior to a greater degree. ThO stands out among the different strategies to impart improved wear resistance, as the approach has previously been used to strengthen the tribo-behavior of Ti and its alloys [2–4]. There is a markedly unusual contrast between the position played by ThO and tribo-oxidation (TO) during counter-body interactions [5]. To facilitate rapid oxidation of the material, exposure of Ti to the atmosphere at elevated temperatures can be established, resulting in the formation of a non-adherent and dense oxide layer on the diffusion-hardened subsurface of the substrate [6,7]. In addition, on Ti64 surfaces, the ThO process will create hard surface layers [8]. The wear resistance of the material components was stated to be increased by the technique as a result [9]. However, to establish the necessary microstructure, the procedure has to be parametrically controlled. The sustained application of high temperatures can contribute to stratification and debonding of the interface, whereas the forming of discontinuous oxide layers is attributable to inadequate temperature conditions applied for long durations [1,10,11]. In contrast to the as-received specimens, due to the oxide layer growth mechanism, the surface roughness of the ThO handled surfaces is observed to have increased. It is the consequence of the development of a porous and stratified structure, formed over time under the conditions of enforced treatment [10,12]. The ThO procedure is usually carried out in the air within a temperature range of 873–923 K for a period of 60–65 hours [12,13]. As a consequence, thin oxide scales (2–3 μm) with a diffused subsurface region (20 μm) beneath it are formed over the usable material surface. The formation of the $TiO_2$ secondary hard phase coupled with the oxygen diffused sub-zone imparts high hardness to the surface of the material, resulting in a corresponding wear reduction. In addition, the intrinsic corrosion-resistant feature of the primary material is preserved following the ThO phase [12,14].

#### 8.2.1.2 Tribo-Oxidation/Oxides (TO) and Tribo-Oxide Layer (TOL)

During the dry sliding interaction of Ti64 with different counter-bodies, microstructural and tribo-behaviorally different layers were found to have formed. Li et al. [15] stated that during the tribo-interaction of Ti64 with AISI 52100 steel, three different zones formed on the substrate surface. The top, intermediate, and bottom layers are (1) tribo-layer (TL), (2) plastically deformed layer (PDL), and (3) base material matrix, respectively. The layer morphology differed with the rise in sliding velocities in such a way that particular properties were correlated with the TL and PDL. The former showed uncompacted (for low and medium speed) and compacted shapes (for high speed) at sliding velocities of 0.75, 2.68, and 4 m/s, respectively. Nevertheless, the size of the latter varied from being thick to thicker and then thin, with increment in velocity.

Pauschitz [16] stated that the wear characteristics associated with dry sliding in metals differ according to the stability and nature of TL. The formation depends on many variables, such as the type, condition, and state of the sliding contact, the base material characteristics, and the counter-face. TLs were narrowly divided into three categories: (1) material migration layer; with a low oxygen content percentage, formed where the counter-face is softer than the material being tested. The migration layer has a structure close to that of the counter-face and normally forms at room temperature; (2) mechanically mixed layer (MML); lower oxygen content levels with better hardness characteristics are available. The MML grows during the interaction of tribo-pairs with mild hardness at comparatively peak temperatures; and (3) composite layer; high oxygen content levels. The nature of the layer, formed at high temperatures and prone to material degradation by wear, is hard and brittle [16].

#### 8.2.1.3 Adiabatic Shear Banding (ASB)

ASB can occur under temperature conditions of high and low pressures. The material's poor thermal conductivity contributes to the creation of shear bands along the grain boundary neighborhoods [17]. It was found in a normal tribological test at flash temperatures [18]. The Ti64 specimens were observed to form ASBs at a strain rate of 0.001 m/s, with an increase in temperature (up to 523 K). In addition, at low flow stress and high tensile circumferential stresses in a narrow zone and the bulge surface equator, respectively, there is a risk for fracturing along the ASB [19]. In the work of Chelliah and Kailas [20], TO, SRR, and ASB have been proposed to be the dominant processes regulating the tribo-behavior of CP Ti. In comparison, at low sliding speeds, the potency of ASB was dramatically high and decreased with increases in the latter. It can then control the resulting high and low wear rates of the material under the conditions described above. The development of ASB may be pronounced in the case of Ti64 at a strain rate equivalent to the sliding speed of the pin at 0.1 m/s [21]. This is due to the fact that the difference in wear rate at the defined low velocity is controlled predominantly by the correspondingly applied strain rate infused microstructural changes produced by the alloy. The ASB phenomenon and the related processes that contributed to the softening of the substance triggered the same thing [22–25].

#### 8.2.1.4 Shear Rate Sensitivity/Response (SRS/SRR)

The imposed strain rate and temperature conditions control the microstructural characteristics of a material, hence the importance of SRR/SRS. Rigney [26] stated that the sliding wear of the materials can be controlled by the microstructure of the near-surface regions. Due to the alteration in the microstructure caused by frictional heating (FH) or the difference in the applied strain rate, a dramatic shift in wear mechanisms may occur. Centered on the changes in microstructure under the applied pressure, strain intensity, and temperature conditions, the SRR method has previously been attempted to test the wear phenomenon [22,23,27].

#### 8.2.1.5 Dynamic Recrystallization (DRX)

Ding et al. [28] stated that the driving force needed for the phase transition is substantially greater than that sufficient for DRX to occur. For Ti-alloys, therefore, it is the former that is very difficult to occur relative to the latter. In addition, during hot operating cycles, phase transition should occur before or after the completion of DRX. Although at a high temperature above the β-transus of 1050°C, only scarce levels of DRX do occur. The phase transition will occur concurrently with the mechanical transformation in the $\alpha+\beta$ phase region. Nonetheless, the DRX remained inactive during hot work in the stated temperature range. In addition, the accumulated energy due to dislocation density for metallic materials adds to the driving forces critical for recrystallization and has a value range of 0.01–0.1 $kJmol^{-1}$. In contrast, the free energy distinction of around 1 $kJmol^{-1}$ of the different phases leads to the driving forces that trigger the transition of the solid phase [29].

#### 8.2.1.6 Flow Stress (FS)

It is stated that the material's FS is primarily governed by the temperature rather than the strain rate [20]. A decremental tendency related to the FS of the Ti64 alloy with an increase in temperature has been observed in the past [18,30,31]. The research by Long and Rack [32] explored the effect of the FS characteristic on the frictional behavior associated with different Ti-alloys, viz. Ti-35Nb-8Zr-5Ta (TNZT and TNZTO), metastable-β (21 SRX), and Ti64. The difference in velocity controls the strain rate encountered by the mating surfaces during reciprocated sliding periods. In addition, the studied surfaces suffer strain deformation and material damage during rapid interactions. The difference in frictional behavior subsequently depends on the alloys' resulting deformation and fracture response. The dynamic frictional coefficient, due to the decreased FS sensitivity of the Ti64 alloy with the applied strain rate, was indicated to be independent of the influence of the transition in sliding velocities [33]. On the other hand, the initial decline in the complex frictional coefficient was attributed to the rise in FS sensitivity and fracture intensity at low contact stress for the reminder. The increase in SRS was previously documented as the indigenous nature of low-strength alloys (BCC) [15,34].

#### 8.2.1.7 External Heating/Heat Flux (EH)

According to Mao et al. [35], the wearing activity of the Ti64 alloy at elevated temperatures remains affected by the loading conditions. At 400°C–500°C, superior tribological characteristics were reported; however, material removal rapidly increased above 200N. In addition, the alloy's low temperature (25°C–200°C) behavior produces peak wear with a rise in the applied load (50–250N). The presence/absence of excess oxides governs the protective action imparted by the formation and existence of TL (MML). With the rise in temperature, the proclivity for oxide/oxide layer formation has been increased. FH should either be assumed to have a marginal impact on the existence of EH, or its subsequent effect is not strongly manifested (engulfed by the former).

#### 8.2.1.8 Frictional Heating (FH)

FH significantly influences the tribological reaction of the Ti64 alloy, as it affects the different mechanisms, viz. TO, SRS, MML, and so on [21]. Sliding velocity and normal load are the main process parameters that govern the variance in FH [36,37]. Consequently, at room temperature, they control the dry sliding wear behavior of the components. Due to the resulting FH-mediated rise in temperature by surface asperity interaction, the tendency to form TO is increased with sliding velocity [15].

#### 8.2.1.9 Thermal Softening (TS)

The results of previous researchers demonstrate the intrinsic characteristics of Ti-alloys to undergo TS at high temperatures [21]. TS of the alloy greatly led to the increased wear rate for the Ti64 slid against equivalent counter-body (at elevated velocities). The fundamental explanation is that the yield strength of the alloy declines with an increase in temperature from room temperature to 200°C to 500°C at the rate of 950 to 650 to 450MPa, respectively [38]. Hence, TS is considered to assist the material removal through the delamination process at the elevated interfacial temperature of the mating surfaces due to FH or rise in ambient temperature.

### 8.2.2 Friction and Wear Characteristics at Interactive Instances

Under combined low load (50–100N) and temperature (25°C–200°C) conditions, the function of TOL is found to be negligible. Gradually, the removal of materials guided by the adhesion mechanism shifted to abrasion and delamination. But many scholars have reported the formation of thin layers during such a state of sliding, contradicting results obtained by a few proposals against it. Conversely, in order to impart protective action, TOL was identified to have considerable thickness and compactness at high load (100–200N) coupled with high-temperature (200°C–500°C) conditions.

The only known mechanism causing surface damage was oxidative wear during tribo-interaction [39].

With ST, quenching, and aging, the integrity of the Ti64 surfaces in the sense of tribo-behavior was found to degrade. The underlying explanations include: (1) HCP structured materials with poor work hardness, (2) phase change leading to the change from ductile to more brittle surface characteristics, and (3) stresses or strains caused by the transition. Micro-plowing, grooving, and wedge forming for the former and micro-cutting aided micro-chip creation for the latter are the signature wear behaviors associated with the as-received and quench-aged Ti64 samples [39].

Under gradual sliding conditions, the change in the wear process is stated to be independent of the differentiation in microstructures (lamellar, bi-modal, or equiaxed) specific to the Ti64 specimen. The general tendency is that oxidative wear dominates the method of material removal at a low sliding speed (0.3 m/s). With the rise in sliding speed (0.6 m/s), there is a change from oxidative to delamination in the wear process. In comparison, metallic delamination dominates the counter-body contact aided wear at peak velocity interaction of the pin (0.9 m/s) [39].

Finally, in terms of the form, composition, and percentage of the generated surface oxides (TiO, $TiO_2$, $Ti_8O_{15}$, $V_2O_3$, and $Fe_2O_3$), the presence and magnitude of the different wear mechanisms (oxidative, delamination, or composite) and the structure of the TL (stability, thickness, and compactness) under distinct sliding conditions, the use of separate counter-body surfaces affects the alloy's tribo-behavior. However, the general tendency was that at low-velocity, low-load, and high-temperature contacts, oxidative wear prevailed and delamination wear prevailed under conditions of medium velocity sliding, peak load, and low temperature. Composite wear (a mixture of oxidative and delamination wear) showed an effective role for all other sliding states [39].

## 8.3 ELECTRICAL DISCHARGE MACHINING ASSISTED ALLOYING

EDM is now a widely used non-conventional machining process. It is the process of machining electrically conductive materials with the aid of precisely controlled sparks that are developed between anode and cathode in the presence of a dielectric medium. Even though it is termed as a "non-conventional" machining process, many EDM tool makers claim that EDM is now the fourth most popular machining method. There are different variants of EDM, viz. Wire EDM, Ram EDM, EDM drilling, and so on. The process has the ability to machine conductive materials irrespective of their hardness (e.g. super alloys, titanium alloys, and so on) to an accuracy of up to 1000th of a millimeter with no mechanical force. The material is removed by a series of rapidly occurring repeated electric arcing discharges between an electrode and the workpiece, in the presence of a dielectric fluid. Figure 8.1 shows a schematic diagram of the EDM process.

With a certain spark gap (usually in the region of 10–80 μm), an NC program helps direct the cutting tool along the desired path. A number of micro-craters on the surface of the workpiece are created by these repeated sparks and the material removal takes place by melting and vaporization along the cutting line. EDM is one of the primary developments in mold and tool making because of these properties.

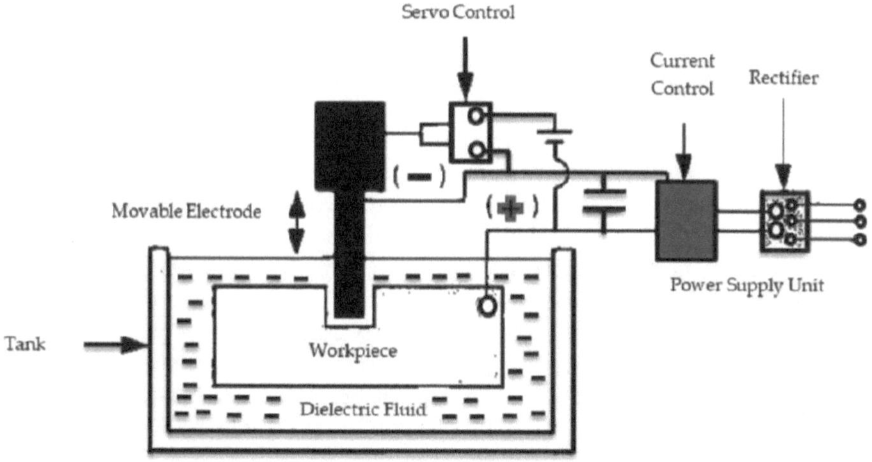


**FIGURE 8.1** Schematic diagram of the EDM process [40]. Reproduced in accordance with Creative Commons Attribution License from Ref. [40].

EDM varies from most chip-making machining procedures where the electrode for material removal does not make direct contact with the workpiece. EDM exerts negligible tool force due to its non-contact existence. The electrode must always be spaced away from the workpiece by the distance (known as a spark gap) needed for sparking. The sparking will stop as the electrode moves into contact with the workpiece, and thereon no material will be removed.

EDM is a thermal process where the material is removed by heat. Heat is produced by the motion of electricity between the electrode and the workpiece, in the form of a spark. Materials are heated to the point that the flame originates and ends, where the material vaporizes at the closest points between the electrode and the workpiece. Nevertheless, only one spark happens at any time. Sparking happens from 2000 to 500,000 sparks per second in a frequency spectrum, making it appear as multiple sparks occurring concurrently. The sparks move from one point on the electrode to another during the EDM process when discharging takes place. This triggers the next spark to appear between the electrode and the workpiece at the next nearest location. The heated area, vaporized material, electrode, and workpiece surfaces are easily cooled off ensuing each spark by the dielectric fluid. The metallurgical changes can, however, arise by the heating of the workpiece surface by the flame. The sparking distance between the electrode and the workpiece requires a dielectric medium to maintain it. Usually, this dielectric medium is fluid. In general, hydrocarbon oil is used by die-sinker style EDM machines, while wire electrical discharge machining (WEDM) machines use deionized water. Until enough electrical voltage is applied to allow it to transform into an electrical conductor, the dielectric fluid acts as an electrical insulator. The dielectric fluid deionizes at the pulse-off time, and the fluid becomes an electrical insulator again. EDM machines use dielectric fluid to monitor the spark gap, cool the heated material to form the EDM chip, and remove EDM debris from the sparking field.

Inside a column of ionized dielectric fluid, a spark develops. A tiny amount of the electrode and workpiece material is vaporized as each spark occurs. In what can be characterized as a cloud, the vaporized substance is located in the spark gap. The vaporized cloud solidifies as the flame is turned off. Each spark then creates an EDM chip or a very small hollow sphere of material made up of tiny chunks of extracts from the electrode and workpiece. The EDM chip has to be separated from the sparking area for successful machining. By flowing dielectric fluid through the sparking gap, elimination of this chip is achieved. The product of small craters being formed by the spontaneous impacts of thousands of sparks is the finish that is created by the EDM process. After each cycle, a new high point forms the rim of the produced crater, making it a possible target for a new cycle. The chips produced by the EDM spark are only 2 μm in size. Overlapping micro-craters are produced due to this circumstance, allowing for the random features on the surfaces developed through EDM. The surface finish is one of EDM's appealing features in many industries, even though the process is relatively slow.

EDM is a potential method capable of machining Ti and its alloys regardless of their production routes, mechanical properties, and chemical properties. The technique is widely utilized for the formation/development of complex shapes and parts exclusively concerning electrically conductive materials but is also compatible with a few non-conductive materials subject to formidable constraints. One of the widely known variants of the technique, viz. WEDM, has the potential for precision machining of small components and tools. The method holds the capacity to form oxygen-rich layers on distinct material surfaces. The incorporation of suspended particulates/powders in the dielectric fluid for EDM, generally referred to as the powder mixed electrical discharge machining (PMEDM) process, can aid in the homogenization of the discharge effect throughout the work material surface. Besides, they assist in the formation of various surface compounds through the phenomenon of material migration. The potential of EDM/WEDM/PMEDM to act as a surface modification technique (SMT) is getting enormous attention to improve the mechanical properties, tribo-response, and bio-degradability of various materials. The approach has a leading edge over other similar SMTs due to its capabilities such as (1) the non-requirement/non-essentiality of pre/post-processing of material surfaces; (2) development of various surface compounds (oxides/carbides) imparting unique properties, viz. biocompatibility and hydrophilicity; (3) formation of nano-porous structures; (4) accretion of hard layers/coatings; and (5) development of wear and corrosion-resistant surfaces. The surface so formed is composed of many randomly distributed overlapping craters. Besides, a positive skewness is imparted to material surfaces by the EDM process through the existence of interspersed peaks and valleys. During tribo-interactions, the vacant space within such craters may entrap the debris particles emanating from the mating surfaces (during dry surface interactions) and entrain lubricants (during wet surface interactions), thereby the effect of friction/wear gets subdued.

### 8.3.1 Mechanism

The application of voltage pulses allows the dielectric in the discharge channel to break down. This results from the displacement of all the electrons released from the

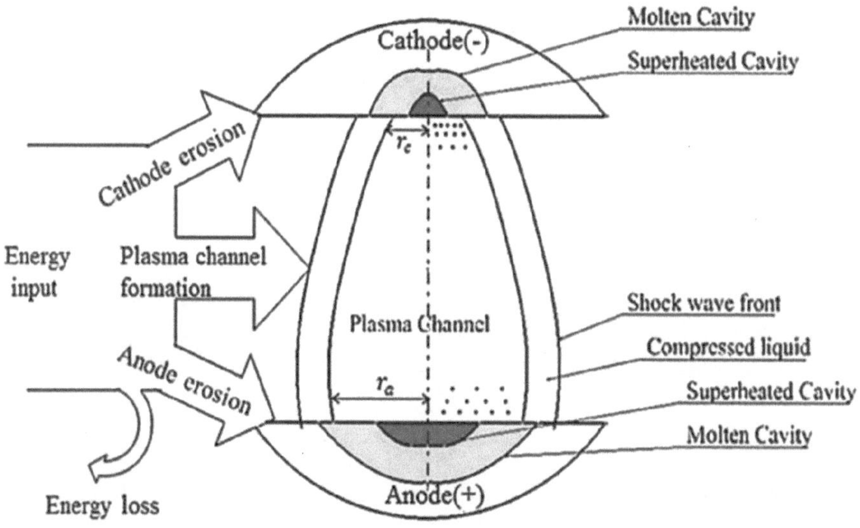


**FIGURE 8.2** Schematic diagram of the basic mechanism involved in the EDM process [41]. (Reproduced with permission from Ref. [41]. Copyright Springer (2015).)

cathode and the stray electrons present in the inter-electrode gap toward the anode by the applied field. These electrons collide with dielectric neutral atoms, forming positive ions and additional electrons that are in turn accelerated toward the cathode and anode, respectively. The schematic diagram of the mechanism of EDM is shown in Figure 8.2.

When the electrons and the positive ions reach the anode and cathode, respectively, they give up their kinetic energy in the form of heat. Temperatures of approximately 8000°C–12,000°C and heat fluxes of up to 1017 W/m$^2$ are thus obtained. For a very short-lasting spark of usually between 0.1 and 2000 μs, the temperature of the electrodes can be increased locally to higher than their normal melting points. The pressure on the plasma channel increases exponentially to levels of up to 200 atmospheres due to the evaporation of the dielectric. Evaporation of the superheated metal is avoided by such great pressures. The pressure drops abruptly at the end of the pulse and the superheated metal evaporates explosively. Thus, the metal is removed from the workpiece. Figure 8.3 indicates multiple stages involved in the EDM process during sparking. A charged electrode is brought close to the piece of work. An insulating oil is found between the electrodes, known as dielectric fluid in EDM. The insulating properties of the dielectric fluid begin to decrease in a narrow channel based on the strongest part of the field as the number of ionic (charged) particles increases. While the voltage reaches its peak, the current remains zero. Nevertheless, as the fluid becomes less of an insulator, the value of the latter increases. The voltage then continues to decline. As the current rises and the voltage begins to decrease, heat builds up rapidly. Some of the dielectric fluid, the work piece, and the electrode are vaporized by the heat, and a discharge channel begins to form between the electrode and the work piece.

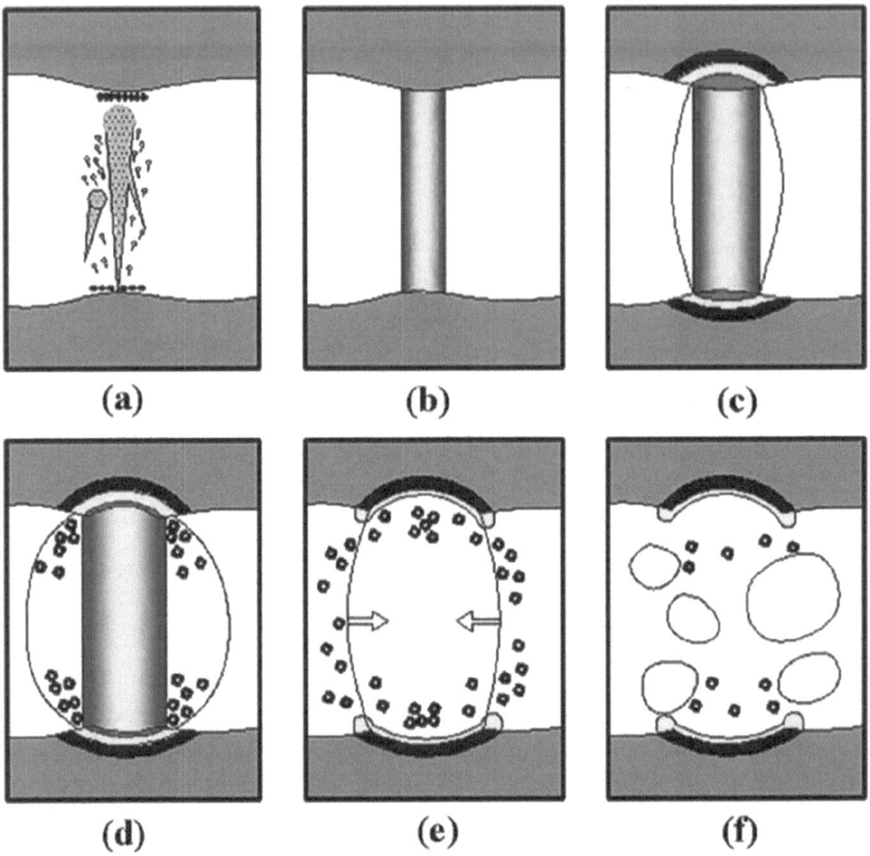

FIGURE 8.3 Illustration of different stages of spark discharges: (a) dielectric breakdown, (b) plasma formation, (c) electrodes melting and vaporization, (d) plasma and bubble extension, (e) plasma collapse and debris ejection, and (f) bubble collapse and deionization [42]. (Reproduced with permission from Ref. [42]. Copyright Elsevier (2016).)

A vapor bubble attempts to expand outward, but the rush of ions into the discharge channel limits its expansion. The highly strong electromagnetic field that has built up attracts these ions. The current continues to increase and the voltage decreases. By the end of Spark-ON time, the current and voltage have settled, heat and pressure have hit their limit inside the vapor bubble, and some metal is being removed. The metal surface immediately under the discharge column is in a liquid state but is kept in place by the vapor bubble pressure. The discharge channel now consists of a superheated plasma made up of vaporized metal, dielectric fluid, and carbon with an intense current flowing through it. The current and voltage reduce to zero at the beginning of off-time. The temperature drops quickly, allowing the vapor bubble to collapse and to remove the molten metal from the workpiece. Fresh dielectric fluid floods in, flushing away the particles and quenching the work piece surface. The molten metal that is not expelled solidifies to form what is regarded as the RL. The ejected metal, along with traces of carbon from the electrode, solidifies into small spheres scattered in the dielectric liquid. The vapor that persists rises to the surface. Without adequate off-time, debris will accumulate, leaving the discharge to be unstable. A DC arc that can damage the electrode and the work piece may be created by this circumstance.

### 8.3.2 Formation of Zones/Layers

The EDM process alters not only the working metal's surface, but also the subsurface. On top of the unaffected working metal, three layers (a total thickness of 0.05–0.1 mm) are formed (Figure 8.4a and b). The different layers developed during the EDM process include the spattered surface layer, the recast (white) layer (2–50 μm), and the heat-affected region (25 μm).

The EDM surface layer is formed when the ejected molten metal and small quantities of electrode material form spheres, and the surface of the work metal is spattered. Through polishing, this spattered material is readily removed. The next layer is the recast (white) layer. EDM process behavior has essentially altered the metallurgical composition and properties of the workpiece in the RL. The liquid metal solidifying in the crater that was not ejected during the pulse-off period forms this layer. The molten metal is quenched by the dielectric rapidly. In this very hard, brittle layer, microcracks can form. In certain applications, the impact of this layer can cause premature failure of the component if it is too thick or is not reduced or eliminated by polishing. In the RL, microcracks exist which may serve as initiation points for failure (reduced fatigue strength). An easily quenched structure characterizes the RL. The structure is typically fragile and extremely hard (65 HRC). It can be porous with distributed microcracks. Abrasive techniques or shot peening operations may eliminate this. The heat-affected zone is the last layer that underwent heating but was not melted. The reasons for the formation of this zone are heating, cooling, and diffused material. Thermal residual stress and grain boundary fractures can be present in the heat-affected zone (HAZ). The depth of the RL and the zone influenced by heat is determined by the material's heat sinking potential and the power used for

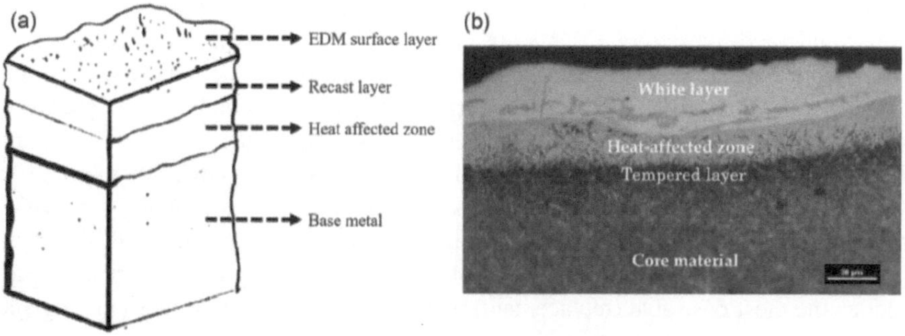


**FIGURE 8.4** Various layers formed during the EDM process: (a) schematic representation, and (b) metallographic image [43,44] ((b) is reproduced in accordance with Creative Commons Attribution License from Ref. [44]).

the cut. The consistency of the surface integrity is affected by this modified metal region. The RL is distinguished by an easily quenched surface, whereas the annealed or tempered structure represents the HAZ.

### 8.3.3 Mechanical, Metallurgical, and Bio-Compatible Properties of the Recast Layer

The solution to toughening mild steel by electric sparking was suggested by Barash and Kahlon [45]. Using a copper (Cu) electrode with paraffin as the dielectric medium, the working material was subjected to erosion. The use of hydrocarbon fluids (to assist the process) resulted in machined surface carburization, forming a superior hardness deposited layer (RL), difficult to extract by post-finishing operations. In addition, by changing the electrical circuit parameters, the material migration/transfer is controlled or subject to control. Tsukahara et al. [46] tested the competency of the Ti hardening using the EDM process. A titanium carbide (TiC) rich crack-free film has been formed over the substrate at low I and PD. Improvement of the material's surface characteristics in terms of hardness, tribo-response, and corrosion resistance is obtained by the processing through EDM. The variations in the surface characteristics of high-speed steel following the EDM method were analyzed by Venkatesh and Parasnis [47]. The deposited coating (RL) over the surface allegedly has qualified properties, i.e., high hardness, elevated thermal stability, and superior wear resistance. The RL got notified to be an alloyed matrix (consisting of carbides) possessing the potential to maintain its hardness characteristics at elevated temperatures. Soni and Chakraverti [48] studied the phenomenon of material migration in EDM during die-steel production (T215Cr12). The process parameters of EDM, viz. I and electrode rotation (RoE), were varied to determine the difference between the characteristics of the surfaces formed over the tool and workpiece. The material from the tool and the workpiece get migrated/transferred to get alloyed to form a deposited matrix (RL) such that the chemical composition of the tool/workpiece is significantly varied.

EDM/PMEDM is an emerging and promising technique capable of developing/imparting bioactive characteristics on implant surfaces [49]. The propensity for the utility and need for artificial joints are increasing substantially. Reportedly, degenerative and inflammatory issues are becoming common in the majority of people aged 40 years (and above). During repetitive cyclic loading conditions, degradation of bones can occur due to the loss of mechanical strength caused by the weakening of bones and cartilages, and inflammation at the joints and synovial membrane. To tackle such severe issues, artificial joints/implants are used to replace the *in-vivo* dysfunctional/load-bearing joints through the total joint replacement (TJR) route [50–52]. Among the various categories of implant materials, Ti64 reserves the top spot as the most desirable (replacement) material for hard-body tissues (e.g. hip and knee joints), due to its superior biocompatibility, excellent material characteristics, and anti-bacterial properties [53]. The two major drawbacks of the various coating techniques used for practical applications are (1) the inability to develop thick layers and (2) incompetent bonding/adhesion of the appendage/deposited layer with the substratum. In the context of bio-implant applications, the body fluid (which is acidic) can easily deteriorate such unstable coatings/layers resulting in premature failure of the implants. Nonetheless, the surface modification using EDM/AMEDM/PMEDM demonstrates the massive potential for the development of bio-adaptive surfaces. The domain is still in the exploratory/experimental stage requiring/demanding extensive attention [49].

Farooq et al. [53] used the PMEDM technique assisted by dielectric dispersed with Si powders to develop a bio-compatible and bioactive surface on the Ti6Al4V (ELI) alloy. The PConc (Si, 5–10–20 g/l), $I$ (5–7–9 A), and $T_{on}$ (50–75–100 µs) got varied to determine the optimized parametric conditions for the formation of a bio-adaptive layer. The surface irregularities/defects (cracks and cavities) produced/developed on PMEDM (processed) surfaces are scarce in number as compared to the conventionally machined surfaces. The PConc (Si) was the dominant factor controlling the SR and RLT. The potential of the PMEDM technique to develop nano-porous (size: 50–200 nm) structures on Ti64 surfaces got revealed through the approach.

## 8.4 SURFACE ALLOYED TI6AL4V

### 8.4.1 Role of the Recast Layer in Controlling Tribological Characteristics

Shunmugam et al. [54] used a powder compact electrode (WC (40%) and Fe (60%)) to improve the wear-resistant characteristics of mild steel through the EDM process. A cutting test was conducted to estimate the wear-response of the coated/uncoated materials. The utilization of materials with high hardness and wear-resistance characteristics in the compact (tool electrode) can cause them to migrate during the EDM process to impart unique properties to the work material surface. Distinct (carbide) phases, viz. $W_2C$, FeC, $Fe_3C$, and $(Fe_3C)H$, were formed over the machined region. A 25%–60% improvement in abrasive wear resistance of mild steel is reportedly achievable through the approach. Wang and Han [55] investigated the competency of coatings (≈550 µm) developed through ESD on Ti6Al4V surfaces. The performance of the appendage layer got evaluated through microstructural analysis, the composition of distinct phases, micro-hardness, and tribo-response (wear behavior). The ESD process led to the formation of an $\alpha'$-phase (martensite) structured layer, which has a strong metallurgical bonding with the substratum. There is a considerable reduction in CoF and mass loss of the material through surface modification, of the order of 0.19 and 2.02 mg, respectively. The hard coating resists the material loss due to adhesion and micro-cutting (abrasion), thereby exhibiting superior wear-resistant characteristics.

### 8.4.2 Surface Response under Dry Sliding Conditions

Tyagi et al. [56] used a green compact tool electrode ($WS_2$ (60%) and Cu (40%)) for the EDM process to improve the tribological performance of mild steel (work material). The variation in the process parameters, viz. I (peak), powder mixing ratio (PMR), and $\tau$ on the output responses, namely morphology, RLT, micro-hardness,

TWR, and tribo-responses got evaluated. The dry nature of the $WS_2$ powder makes it impossible to prepare the solid electrode; this was the motivation behind the utilization of a conductive material (Cu) to manufacture the green compact. A proportionate increment in the thickness of the coating (RLT) got observed with an increment in percentage composition of $WS_2$ in the tool electrode. Optimum parametric conditions for improved performance of the deposited layer in the context of the chosen performance parameters are $I = 7$ A, PMR = $WS_2$ (60%) and Cu (40%), and $\tau = 50\%$. The debris analysis ensuing wear test revealed the formation of various compounds, viz. $Cr_2(SO_4)_3$, $SiO_2$, $Cu(PO_3)_2$, and $WO_3$. The presence of the former two compounds confirms that the wear is more for the disk (EN31) than the pin (coated material). Reportedly, the formation of $WO_3$ assists in imparting lubricating characteristics.

Philip et al. [57] conducted experimental investigations to evaluate the wear behavior of electrical discharge machined Ti64 surfaces under ambient conditions. BTi64 demonstrated peak wear at 300 rpm due to abrasion-assisted delamination. ETi64 surfaces exhibit wear-resistant nature due to the presence of hard and non-etchable RL present over the substrate. Philip et al. [58] determined the optimum parametric conditions for developing tribo-adaptive layers on Ti64 surfaces and investigated the properties of the surface during dry interactive sliding with the counter-face. Under ambient conditions, the RL existent over the ETi64 specimens can impart ceramic characteristics leading to the reduction in wear. The random textures developed over the ETi64 specimens can trap debris particulates (during interactive sliding), assisting in build-up stabilization. Philip et al. [59] tested the tribological competency of the layers (at high-temperatures of 200°C, 400°C, and 600°C) deposited through EDM. At elevated temperatures, distinct oxides, viz. $TiO_2$ and $Ti_8O_{15}$, get developed over the substrate under the applied load conditions. The carbides such as TiC and $Ti_{24}C_{15}$ impart high hardness characteristics to the RL, enabling them to extend protective action (against wear) and stabilize the oxides. Philip et al. [60] incorporated SiC particulates in the dielectric to develop wear-resistant layers on Ti64 through the PMEDM route. ETi64 specimens have better tribological behavior than BTi64 under all sliding conditions. The incorporation of secondary phases in the form of abrasive particles can cause catastrophic damage during sliding under high load conditions.

## 8.5 FUTURE SCOPE

The thrust areas in the subject context for the current/future researchers to focus on are detailed below:

- The RL deposited ensuing the EDM/PMEDM process governs the characteristics of the machined/modified surfaces. The past/previous investigations focus more on the RLT, morphology/topography, and mechanical properties (hardness and toughness) of the layer. The least explored characteristics of the RL include metallurgical behavior (phase distribution/transformation and grain size/refinement), bond strength (coherence with the substratum), and evaluation of the formation/stability/distribution/impact of the various surface compounds.

- Various researchers have attempted to model/simulate the EDM process through mathematical/numerical/analytical routes; nevertheless, they are unable to capture the true essence of the process due to the stochastic nature of the technique.
- The competency of the EDM/PMEDM technique for texturing applications is a positive possibility requiring precise/immediate attention. Since the process can develop undulations through the formation of craters (whose size can be controlled by varying the process parameters) over a large area, they can be tested for surfaces undergoing interactive sliding with distinct counter-bodies. The random textures can encapsulate debris (leading to build-up stabilization) and lubricants (assisting prolonged lubrication) during dry and wet interactive sliding, respectively.
- Improving the tribological and bio-adaptive characteristics of material surfaces through EDM/PMEDM is still in the exploratory stage. Advanced analysis/investigations are essential to comprehend the responses of thereby developed surfaces under *in-vivo/in-situ* conditions for implementation in practical applications.

## 8.6 CONCLUSIONS

The tribo-characteristic evaluations of Ti64 (unmachined and EDM machined) critically compared were found to produce promising results. The material in its bare form showed catastrophic material damage resulting in severe damage and seizure on interaction with counter-body surfaces. Now, a material whether it possesses any much of high-end material capabilities or not, if unable to sustain to be stable enough during material to material interaction, can be downgrading. All the mechanical systems irrespective of their complexity and precise significance have their components undergoing tribo-interactions. Such asperity mating conditions lead to failure, aided by the various wear mechanisms of adhesion, abrasion, erosion, corrosion, and so on during the long run. These damage-infusing phenomena are the prime causes of minor/severe breakdowns of machinery and advanced production systems, across the globe. Such mechanisms should be degraded from their roots to prevent the occurrence of adverse scenarios. Among the aforementioned, abrasion, adhesion, and delamination are significant contributors to the material failure of Ti64 in practical applications.

The real-time experimentation and testing of Ti64 mating against hard counter-bodies were observed to produce varying characteristics under ambient-, low-, and high-temperature conditions. The degree of damage suffered by bare Ti64 was severe in its intensity for the same process conditions. The RL protected Ti64 surfaces showed sustained credibly, for the layer acted like a hard and suffice to prevent damage coating. The frictional and wear characteristic plots were found to compliment this argument. Under critical conditions of high load, high temperature, and high speed, the results were observed to be similar due to spalling/peeling-off effect of RL coatings under such adverse damaging conditions. However, the generation of the HAZ in the subzones as a result of EDM processing resulted in phase or metallurgical change, which plays a vital role in delaying severe damage, on exposure of a bare material. This was found to be a critical factor in suppressing the TS behavior of Ti64 at elevated temperatures.

The SMTs separately implemented after machining can be uneconomic, time-consuming, and impractical in many ways. So, developing processes/technologies which can aid in achieving the former and the latter, simultaneously (in one go) will have the added advantage for its implementation in advanced/specialized applications. The RL formed ensuing the EDM process if subjected to further research and investigation can be significant in improving the performance of the so-formed surfaces as well as the development of new surface modification processes. Another added advantage is the elimination of the post-processing methods that have to be implemented to remove the RL formed by the process. The argument that the layer formed can act as a protective modified coating is valid in itself due to its observable improved properties that are already discussed. Moreover, achieving coherent tribo-characteristics is a must for advanced applicability, so the development of a tangible method to improve the friction and wear properties of high strength and hard-to-cut materials like Ti64 is significant. Such a positive possibility, with due consideration of the downsides, was hitherto disregarded and is extremely vital to be elucidated.

## REFERENCES


1. Siva Rama Krishna, D., Brama, Y. L., and Sun, Y., 2007, "Thick Rutile Layer on Titanium for Tribological Applications," *Tribol. Int.*, **40**(2), pp. 329–334.
2. Dong, H., and Bell, T., 2000, "Enhanced Wear Resistance of Titanium Surfaces by a New Thermal Oxidation Treatment," *Wear*, **238**(2), pp. 131–137.
3. Dong, H., and Li, X. Y., 2000, "Oxygen Boost Diffusion for the Deep-Case Hardening of Titanium Alloys," *Mater. Sci. Eng. A*, **280**(2), pp. 303–310.
4. Yazdanian, M. M., Edrisy, A., and Alpas, A. T., 2007, "Vacuum Sliding Behaviour of Thermally Oxidized Ti–6Al–4V Alloy," *Surf. Coatings Technol.*, **202**(4–7), pp. 1182–1188.
5. Budinski, K. G., 1991, "Tribological Properties of Titanium Alloys," *Wear*, **151**(2), pp. 203–217.
6. Frangini, S., Mignone, A., and de Riccardis, F., 1994, "Various Aspects of the Air Oxidation Behaviour of a Ti6Al4V Alloy at Temperatures in the Range 600–700°C," *J. Mater. Sci.*, **29**(3), pp. 714–720.
7. Chaze, A. M., and Coddet, C., 1986, "The Role of Nitrogen in the Oxidation Behaviour of Titanium and Some Binary Alloys," *J. Less Common Met.*, **124**(1–2), pp. 73–84.
8. Borgioli, F., Galvanetto, E., Fossati, A., and Pradelli, G., 2004, "Glow-Discharge and Furnace Treatments of Ti-6Al-4V," *Surf. Coatings Technol.*, **184**(2–3), pp. 255–262.
9. Mushiake, M., Asano, K., Miyamura, N., and Nagano, S., 1991, "Development of Titanium Alloy Valve Spring Retainers," *SAE Trans.*, **100**(5), pp. 475–483.
10. Bertrand, G., Jarraya, K., and Chaix, J. M., 1984, "Morphology of Oxide Scales Formed on Titanium," *Oxid. Met.*, **21**(1–2), pp. 1–19.
11. Qin, Y., Lu, W., Zhang, D., Qin, J., and Ji, B., 2005, "Oxidation of in Situ Synthesized TiC Particle-Reinforced Titanium Matrix Composites," *Mater. Sci. Eng. A*, **404**(1–2), pp. 42–48.
12. Dong, H., Bloyce, A., Morton, P. H., and Bell, T., 1997, "Surface Engineering to Improve Tribological Performance of Ti–6Al–4V," *Surf. Eng.*, **13**(5), pp. 402–406.
13. Dearnley, P.., Dahm, K.., and Çimenoğlu, H., 2004, "The Corrosion–Wear Behaviour of Thermally Oxidised CP-Ti and Ti–6Al–4V," *Wear*, **256**(5), pp. 469–479.

14. Güleryüz, H., and Çimenoğlu, H., 2004, "Effect of Thermal Oxidation on Corrosion and Corrosion–Wear Behaviour of a Ti–6Al–4V Alloy," *Biomaterials*, **25**(16), pp. 3325–3333.
15. Li, X. X., Zhou, Y., Ji, X. L., Li, Y. X., and Wang, S. Q., 2015, "Effects of Sliding Velocity on Tribo-Oxides and Wear Behavior of Ti–6Al–4V Alloy," *Tribol. Int.*, **91**, pp. 228–234.
16. Pauschitz, A., Roy, M., and Franek, F., 2008, "Mechanisms of Sliding Wear of Metals and Alloys at Elevated Temperatures," *Tribol. Int.*, **41**(7), pp. 584–602.
17. Nemat-Nasser, S., Guo, W.-G., Nesterenko, V. F., Indrakanti, S. S., and Gu, Y.-B., 2001, "Dynamic Response of Conventional and Hot Isostatically Pressed Ti–6Al–4V Alloys: Experiments and Modeling," *Mech. Mater.*, **33**(8), pp. 425–439.
18. Rittel, D., and Wang, Z. G., 2008, "Thermo-Mechanical Aspects of Adiabatic Shear Failure of AM50 and Ti6Al4V Alloys," *Mech. Mater.*, **40**(8), pp. 629–635.
19. Kailas, S. V., Prasad, Y. V. R. K., and Biswas, S. K., 1994, "Flow Instabilities and Fracture in Ti-6Al-4V Deformed in Compression at 298 K to 673 K," *Metall. Mater. Trans. A*, **25**(10), pp. 2173–2179.
20. Chelliah, N., and Kailas, S. V., 2009, "Synergy between Tribo-Oxidation and Strain Rate Response on Governing the Dry Sliding Wear Behavior of Titanium," *Wear*, **266** (7–8), pp. 704–712.
21. Pottirayil, A., and Kailas, S. V., 2017, "Dry Sliding Wear Behavior of Ti-6Al-4V Pin Against SS316L Disk at Constant Contact Pressure," *J. Tribol.*, **139**(2), p. 021603.
22. Kailas, S. V., and Biswas, S. K., 1999, "Sliding Wear of Copper Against Alumina," *J. Tribol.*, **121**(4), pp. 795–801.
23. Kailas, S. V., and Biswas, S. K., 1995, "The Role of Strain Rate Response in Plane Strain Abrasion of Metals," *Wear*, **181–183**, pp. 648–657.
24. Ramirez, A. C., 2008, *Microstructural Properties Associated with Adiabatic Shear Bands in Titanium-Aluminum-Vanadium Deformed by Ballistic Impact*, The University of Texas at El Paso.
25. Me-Bar, Y., and Shechtman, D., 1983, "On the Adiabatic Shear of Ti–6Al–4V Ballistic Targets," *Mater. Sci. Eng.*, **58**(2), pp. 181–188.
26. Rigney, D., 1998, "Microstructural Evolution during Sliding," *Wear of Engineering Materials*, pp. 3–12.
27. Biswas, S. K., and Kailas, S. V., 1997, "Strain Rate Response and Wear of Metals," *Tribol. Int.*, **30**(5), pp. 369–375.
28. Ding, R., Guo, Z. X., and Wilson, A., 2002, "Microstructural Evolution of a Ti–6Al–4V Alloy during Thermomechanical Processing," *Mater. Sci. Eng. A*, **327**(2), pp. 233–245.
29. Humphreys, F. J., and Hatherly, M., 2012, *Recrystallization and Related Annealing Phenomena*, Elsevier, Oxford, UK.
30. Osovski, S., Rittel, D., and Venkert, A., 2013, "The Respective Influence of Microstructural and Thermal Softening on Adiabatic Shear Localization," *Mech. Mater.*, **56**, pp. 11–22.
31. Johnson, G. R., 1983, "A Constitutive Model and Data for Materials Subjected to Large Strains, High Strain Rates, and High Temperatures," In *Proceedings of 7th International Symposium on Ballistics*, The Hague, The Netherlands, pp. 541–547.
32. Long, M., and Rack, H. J., 2001, "Friction and Surface Behavior of Selected Titanium Alloys during Reciprocating-Sliding Motion," *Wear*, **249**(1–2), pp. 157–167.
33. Biswas, C. P., 1973, "Strain Hardening of Titanium by Severe Plastic Deformation," Massachusetts Institute of Technology.
34. Laird, C., 1982, "Strain Rate Sensitivity Effects in Cyclic Deformation and Fatigue Crack," In *Proceedings of 1st International Conference on Corrosion Fatigue up to Ultrasonic Frequencies*, The Met. Soc. of AIME, Pennsylvania, PA.
35. Mao, Y. S., Wang, L., Chen, K. M., Wang, S. Q., and Cui, X. H., 2013, "Tribo-Layer and Its Role in Dry Sliding Wear of Ti–6Al–4V Alloy," *Wear*, **297**(1–2), pp. 1032–1039.

36. Cui, X. H., Mao, Y. S., Wei, M. X., and Wang, S. Q., 2012, "Wear Characteristics of Ti-6Al-4V Alloy at 20–400°C," *Tribol. Trans.*, **55**(2), pp. 185–190.
37. Ming, Q., Yongzhen, Z., Jun, Z., and Jianheng, Y., 2006, "Correlation between the Characteristics of the Thermo-Mechanical Mixed Layer and Wear Behaviour of Ti–6Al–4V Alloy," *Tribol. Lett.*, **22**(3), pp. 227–231.
38. Collings, E. W., 1984, *The Physical Metallurgy of Titanium Alloys*, American Society for Metals, Metals Park Ohio.
39. Philip, J. T., Mathew, J., and Kuriachen, B., 2019, "Tribology of Ti6Al4V: A Review," *Friction*, **7**(6), pp. 497–536.
40. Abdudeen, A., Abu Qudeiri, J. E., Kareem, A., Ahammed, T., and Ziout, A., 2020, "Recent Advances and Perceptive Insights into Powder-Mixed Dielectric Fluid of EDM," *Micromachines*, **11**(8), p. 754.
41. Kuriachen, B., Varghese, A., Somashekhar, K. P., Panda, S., and Mathew, J., 2015, "Three-Dimensional Numerical Simulation of Microelectric Discharge Machining of Ti-6Al-4V," *Int. J. Adv. Manuf. Technol.*, **79**(1–4), pp. 147–160.
42. Liu, Q., Zhang, Q., Zhang, M., and Zhang, J., 2016, "Review of Size Effects in Micro Electrical Discharge Machining," *Precis. Eng.*, **44**, pp. 29–40.
43. Handa, V., Goyal, P., and Kumar, R., 2020, "Review on Surface Modifications of the Workpiece by Electric Discharge Machining," pp. 21–29.
44. Świercz, R., Oniszczuk-Świercz, D., and Chmielewski, T., 2019, "Multi-Response Optimization of Electrical Discharge Machining Using the Desirability Function," *Micromachines*, **10**(1), p. 72.
45. Barash, M. M., and Kahlon, C. S., 1964, "Experiments with Electric Spark Toughening," *Int. J. Mach. Tool Des. Res.*, **4**(1), pp. 1–8.
46. Tsukahara, H., Sone, T., and Masui, K., 2000, "Surface Hardening of Titanium Using EDM Process," *Titan. Japan(Japan)*, **48**(2), pp. 47–49.
47. Venkatesh, V. C., and Parasnis, S., 1972, "Surface Transformation in High Speed Steel after Electro Discharge Machining," In *Proceedings of the 5th AIMTDR Conference*, IIT, Roorkee, India, pp. 639–649.
48. Soni, J. S., and Chakraverti, G., 1996, "Experimental Investigation on Migration of Material during EDM of Die Steel (T215 Cr12)," *J. Mater. Process. Technol.*, **56**(1–4), pp. 439–451.
49. Aliyu, A. A., Abdul-Rani, A. M., Ginta, T. L., Prakash, C., Axinte, E., Razak, M. A., and Ali, S., 2017, "A Review of Additive Mixed-Electric Discharge Machining: Current Status and Future Perspectives for Surface Modification of Biomedical Implants," *Adv. Mater. Sci. Eng.*, **2017**, pp. 1–23.
50. Li, Y., Yang, C., Zhao, H., Qu, S., Li, X., and Li, Y., 2014, "New Developments of Ti-Based Alloys for Biomedical Applications," *Materials (Basel)*, **7**(3), pp. 1709–1800.
51. Geetha, M., Singh, A. K., Asokamani, R., and Gogia, A. K., 2009, "Ti Based Biomaterials, the Ultimate Choice for Orthopaedic Implants – A Review," *Prog. Mater. Sci.*, **54**(3), pp. 397–425.
52. Prakash, C., Kansal, H. K., Pabla, B., Puri, S., and Aggarwal, A., 2016, "Electric Discharge Machining – A Potential Choice for Surface Modification of Metallic Implants for Orthopedic Applications: A Review," *Proc. Inst. Mech. Eng. Part B J. Eng. Manuf.*, **230**(2), pp. 331–353.
53. Umar Farooq, M., Pervez Mughal, M., Ahmed, N., Ahmad Mufti, N., Al-Ahmari, A. M., and He, Y., 2020, "On the Investigation of Surface Integrity of Ti6Al4V ELI Using Si-Mixed Electric Discharge Machining," *Materials (Basel)*, **13**(7), p. 1549.
54. Shunmugam, M. S., Philip, P. K., and Gangadhar, A., 1994, "Improvement of Wear Resistance by EDM with Tungsten Carbide P/M Electrode," *Wear*, **171**(1–2), pp. 1–5.
55. Wang, W., and Han, C., 2018, "Microstructure and Wear Resistance of Ti6Al4V Coating Fabricated by Electro-Spark Deposition," *Metals (Basel)*, **9**(1), p. 23.

56. Tyagi, R., Das, A. K., and Mandal, A., 2018, "Electrical Discharge Coating Using WS2 and Cu Powder Mixture for Solid Lubrication and Enhanced Tribological Performance," *Tribol. Int.*, **120**, pp. 80–92.
57. Philip, J. T., Kumar, D., Mathew, J., and Kuriachen, B., 2019, "Wear Characteristic Evaluation of Electrical Discharge Machined Ti6Al4V Surfaces at Dry Sliding Conditions," *Trans. Indian Inst. Met.*, **72**(10), pp. 2839–2849.
58. Philip, J. T., Kumar, D., Joshi, S. N., Mathew, J., and Kuriachen, B., 2019, "Monitoring of EDM Parameters to Develop Tribo-Adaptive Ti6Al4V Surfaces through Accretion of Alloyed Matrix," *Ind. Lubr. Tribol.*, **72**(3), pp. 291–297.
59. Philip, J. T., Kumar, D., Mathew, J., and Kuriachen, B., 2020, "Experimental Investigations on the Tribological Performance of Electric Discharge Alloyed Ti–6Al–4V at 200–600°C," *J. Tribol.*, **142**(6), p. 061702.
60. Philip, J. T., Kumar, D., Mathew, J., and Kuriachen, B., 2020, "Sliding Behavior of Secondary Phase SiC Embedded Alloyed Layer Doped Ti6Al4V Surfaces Ensuing Electro Discharge Machining," pp. 163–172.

# Index